TORNADOES AND WATERSPOUTS IN IRELAND:
ANCIENT AND MODERN

Tornadoes and Waterspouts in Ireland: Ancient and modern

JOHN TYRRELL

CORK UNIVERSITY PRESS

First published in 2021 by
Cork University Press
Boole Library
University College Cork
CORK
T12 ND89
Ireland

Library of Congress Control Number: 2021932781
Distribution in the USA: Longleaf Services, Chapel Hill, NC, USA

British Library Cataloguing in Publication Data
A CIP record for this book is available from the British Library.

ISBN: 978-1-78205-459-7

Printed in Malta by Gutenberg Press
Print origination & design by Carrigboy Typesetting Services
www.carrigboy.co.uk

COVER IMAGES – A developing vortex near Maghera, Northern Ireland, on 8 June 2011, by permission of Martin McKenna.

www.corkuniversitypress.com

Contents

ACKNOWLEDGEMENTS   vii

LIST OF FIGURES AND TABLES   ix

1   The Quest   1
2   Encounters with Tornadoes   8
3   The Early History of Irish Tornadoes   17
4   A Prisoner of Consensus Science   41
5   The Making of a Tornado   59
6   Waterspouts on Irish Waters   83
7   Tornado Strikes in Ireland: Chronologies and patterns   111
8   Investigating Irish Tornadoes   136
9   Trails of Destruction   168
10   Popular Responses over Past Centuries   194
11   Suddenly Exposed   213
12   An International Context   223

ENDNOTES AND REFERENCES   234

APPENDIX Irish Tornadoes and Waterspouts Referenced
         in the Text   245

BIBLIOGRAPHY   249

INDEX   259

Acknowledgements

In Ireland there is a small community of weather enthusiasts who express their fascination in many different ways. They chase storms, they watch the clouds and note the winds, they measure the weather, they keep diaries, they take remarkable photographs and they have memories full of the wonder they have experienced as the sky has taken on unusual forms and surprised them, despite their (often) long-term experience of the weather. They are of all ages, from primary school age to those in retirement. Some are privileged to be professionally involved in these activities; many are not. But the fascination still consumes them. This volume could not have been produced without them and I am deeply grateful to them all, however brief or expansive our encounter has been.

In many cases our encounters have been on site, where a tornado has forged across the landscape. These enthusiasts have provided support and local insights which brought the tornado event even more alive than just the damage and destruction spread across the fields and woodlands could. They often assisted in providing links to local communities as well as contacts with relatively isolated individuals who had important experiences to share. It was a privilege to hear their stories.

I would particularly like to acknowledge the numerous reports and other inputs of Henry Skeath, David Meskill, Martin McKenna, Martin Sweeney, Martin North and Ian Rippey. In addition, there has been the invaluable help of Met Éireann staff who linked me with so many eyewitnesses who wished to pass on valuable information for the record. Among these staff members were Peter Lennon, Brian Delaney and Ruth Coughlan, as well as academic colleagues Kieran Hickey and Nicholas Betts. I also acknowledge with gratitude the critical stimulation and support of tornado researchers beyond the shores of Ireland.

They included Derek Elsom, Terence Meaden, Tony Gilbert, Paul Knightley, Paul Brown, Robert Doe, Michael Rowe, Peter Kirk, Charles Doswell, Nicolai Dotzek, Michalis Sioutas, Alois Holzer and Pieter Groenemeijer. These are just a few of the many whose contributions have made the research behind this publication possible. It is an incomplete story. It is my hope that this book will encourage and enable more to join the small army of weather observers and weather enthusiasts from all backgrounds, young and old, to continue to observe, record and report details of tornado events in Ireland.

List of Figures and Tables

Figures

1.1	The overturned mobile home near Youghal.	3
1.2	The track of the Youghal tornado.	5
2.1	Lines of sight of various photographers to the Kinrush funnel cloud. (Photo from Kinnegoe by Paddy Prunty)	13
3.1	A dragon in the sky near Ballincollig, County Cork. (Photo by Tony Quane)	23
3.2	A ship's hull with a descending anchor in the sky, being a funnel descending over County Antrim. (Photo by John McConnell)	28
3.3	A vortex appearing like a pillar of smoke, County Waterford. (Photo by Joe Cashin)	33
4.1	The concrete base, supporting blocks and wire cables from which a mobile home was ripped away at Derrynacross.	55
4.2	A field sketch of the funnel cloud made during a storm chase; view is from County Tipperary towards the south.	56
5.1	Sketch of the Stack Hills tornado based on the photo from the IRCG rescue helicopter.	60
5.2	The visible and invisible parts of a tornado (simplified).	62
5.3	A tornado funnel over the Galtymore mountain range, County Tipperary. (Photo by Simon Woodworth)	65
5.4	Cumulus congestus with a funnel cloud in Connemara. (Photo by John Armstrong)	69
5.5	Photo of the Banemore tornado, County Kerry. (Photo by Tim Griffin)	73
5.6	Ingredients that produced the Banemore tornado.	74
5.7	The Ballysadare tornado track with the wind shear directions between the upper air and the ground surface, the funnel's descent and the track of the tornado.	80

6.1   The characteristics of a waterspout as may be seen by an
      observer.   95
6.2   Synoptic characteristics for the Clew Bay and other
      waterspouts, 17 August 2000.   99
6.3   The waterspout that developed suddenly on Lough Neagh,
      June 2014. (Photo by Gemma Convey)   103
6.4   The 'white lady' of Doo Lough. (Photo by Roger Derham)   106
7.1   Annual tornadoes and waterspouts per 1000 km², by county,
      1998–2017.   117
7.2   The geographical divide between the ten counties with the
      highest and lowest number of tornadoes per 1000 km²,
      1998–2017. The individual highest and lowest are in bold.   118
7.3   A funnel cloud near Maghera on 8 June 2011. (Photo by
      Martin McKenna)   123
7.4   Synoptic features for the Carrigallen tornado, County
      Leitrim, 26 January 2002.   124
8.1   The wedge tornado crossing counties Galway and
      Roscommon. (Photo by Laurence Cheeseman)   139
8.2   The tracks of the County Galway–County Roscommon
      tornadoes.   140
8.3   Significant features from the synoptic charts for the
      Westmeath tornadoes, 2001.   142
8.4   The tracks of the County Westmeath tornadoes.   144
8.5   Significant ingredients from the synoptic charts for
      17 August 2001.   148
8.6   The width of the Carrigallen tornado track and the three
      development stages.   150
8.7   Tornado track length (km) frequencies by percentage.   154
8.8   The Ardmore–Tolans Point tornado track.   154
8.9   The frequency of tornado path widths (m), by percentage.   156
8.10 The track of the Togher tornado.   157
8.11 How tornado path width varies with tornado path length.   157
9.1   Impacts of the tornado at Brow Head.   169
9.2   Twisted structural damage due to a tornado at Coolrain.   173
9.3   The Castleisland tornado track and the landing approach
      into Kerry airport.   177

9.4 A boat lifted from its moorings to the back of the beach at
     Broadhaven Bay, County Mayo, August 2005. 180
9.5 The track of the Limerick tornado of 1851. 182
9.6 One of a number of corrugated galvanised iron sheets that
     went missing during a tornado in County Tyrone, but was
     found on a distant farm during harvesting some considerable
     time later. 188
10.1 Lough Fenagh – the site of a possible sixth-century
     waterspout. 199
10.2 Traditional turf cutting that created a working environment
     of deep trenches in the bogs of the west of Ireland. 203
11.1 Sketch of the Bailieborough tornadoes by a child eyewitness.
     Note the damaged rooftops (exposed beams) and felled trees. 222
12.1 Ireland's theoretical location between Tornado Alley and
     those states with fewer tornadoes than Ireland. 225
12.2 USA tornadoes 1997 to 2017 of F3 or greater. 227
12.3 Tornado totals for European states, 1950 to 2015 and
     per 1000 km². 229
12.4 Annual tornado totals for the UK, 1990 to 2010. 231

Tables

4.1 The tornado intensity scale developed by TORRO. 52
7.1. Annual tornado (TN) data, 1998 to 2017. 113
7.2 Monthly distribution of tornadoes and funnel cloud
     characteristics over twenty years. 122
7.3 The most common hours for tornado initiation and the
     earliest and latest times per month (UTC), from 1980.
     Italicised bold times are events that initiated during the
     night. 130
7.4 Probabilities for annual tornado and waterspout totals
     being equal to or greater than given frequencies. 132
8.1 The Carrigallen tornado: Damage details in each stage of
     its development. 150
8.2 Map annotations for field mapping of a tornado track. 162
8.3 Information gathered by eyewitness questionnaires. 164

8.4   Data for the meteorological environment of each tornado
      event.   165
8.5   Contrasts in the damage patterns of the tornado and
      straight-line winds.   166
9.1   Waterspout threats on Irish waters.   179

CHAPTER ONE

The Quest

Go. Observe. Believe!

A visit to a tornado impact site for the first time leaves a profound impression of a reality that, in Ireland, defies expectation. My experience of one such occasion led to the gradual stimulation of a quest for the truth about tornadoes in Ireland; one that would lead me to explore remote peat bogs and seacoasts, urban suburbia and rural landscapes, as well as mountains, lakes and forests.

The event itself took place at 7 a.m. on a cold winter's morning on the outskirts of Youghal in County Cork, in February 1995. It caused a local stir and was reported in the *Irish Examiner* (then the *Cork Examiner*). A check in *The Irish Times,* regarded at the time as the dominant national newspaper, revealed no report at all.

The press report aroused curiosity. But the interpretation presented was unconvincing, so I visited the site four days later. The road into Youghal passed by the Youghal greyhound track. Its wall was partially down and a nearby bus stop was curiously twisted. Both of these features had been noted in the press report. But this did not seem to be very exceptional. Indeed, hundreds of road users must have passed since the storm and thought nothing of the roadside damage. Storm damage is common in Ireland. The report appeared to be correct so far. Next to the wall was a petrol station. What had been lacking in any of the press accounts was an eyewitness account. So, if the petrol station had been open at the time the wall was damaged, there was a good chance that someone had seen what had happened. At first my hopes were raised.

Yes, it had been open and serving customers. However, just before the wall of the greyhound track came down a thunderstorm and torrential rain sent everyone indoors for shelter. They heard the rain thundering on the roof and the howling wind, and the station was severely buffeted. But the entrance faced the opposite direction to the greyhound track, so nothing was seen of what had happened.

A nearby coast road led to two medium sized caravan sites in which structural damage to mobile homes had been reported. A narrow road gave access to the site. Along the road I sought more impressive evidence. Sometimes what you see is so bizarre that the senses distort the reality of what is right before your eyes. Curiously, on the left side of the narrow road stood a single mobile home, all by itself on the edge of a marsh. Just past this, a turn to the right gave access into the caravan park. There the proprietor explained how some of his mobile homes had been wrecked in the early morning in a storm that had arrived from the sea. The damage was confined to a small corner of his campsite within what could be described as a sharply defined track of damage and debris. On one side all the mobile homes were undamaged and undisturbed. On the other, the destruction was extremely severe. It is tempting to say that the damage was total, but that was not the case. Curiously, by the side of a mobile home that had been shattered into many fragments was another that seemed to be relatively intact, although it was a few metres away from the blocks upon which it had rested before the storm. The shock came on closer inspection. The caravan was upside down (Figure 1.1)! It was not on its side, as a severe gale might have bowled it over. The pattern of damage indicated that it had turned right over in the air before it landed on its roof. It now lay on its back with its legs sticking up in the air like a proverbial dead cow.

Any caravan site on the south coast of Ireland is vulnerable to gale force winds, since these are not infrequent. The potential vulnerability of this caravan site to such winds had been identified already by the owner. As a result, the site had been carefully located behind a high bank that bordered the beach and the caravan park. However, the mobile homes most severely damaged were those in the most sheltered location behind the bank. Indeed, the most startling effect of the storm was in this area. Along the line of damaged mobile homes there was a space where one of

1.1 The overturned mobile home near Youghal.

the homes was missing; but not for long. It had been found outside the campsite. In fact, it was the single mobile home I had passed on accessing the campsite gates. Next came the part that was really difficult to grasp. To get to its new location, the home must have been lifted into the air ('levitated' is the formal expression) and carried over the nearby perimeter fence. It certainly did not pass through the fence as it would have brought the whole structure down in the process. Then it was carried across the lane that ran parallel to it before being dumped back down onto the ground. So extreme was the force required to lift such a structure so high and for such a distance that it was quite bewildering to comprehend.

Returning to the main Cork road with a keener sense of anticipation and sharper eyes, I quickly realised that the corner of a second caravan site along the road had also suffered damage to a few of its mobile homes. Beyond that, across a small marsh, the stand of the greyhound track had been caught by the ferocious winds and damaged, including numerous panes of glass. These may have been shattered partly by debris from the caravan sites.

Between the destruction in the caravan sites and the damaged structures of the greyhound track, as far as the Cork road, there was a trail of damage that formed an almost perfect straight line. The levitation of the mobile homes and the sharp damage boundary was further evidence. Even though there were no eyewitnesses to the event, how much stronger did the evidence have to be to conclude that a tornado was responsible for all this?

There could have been a human tragedy as well, but this was mercifully avoided. The inflow of the tornado winds speeding around its central funnel, estimated from the damage effects at about 193 km/h,[1] would have been extremely dangerous to anyone caught in the open. Anyone waiting for an early morning bus (no doubt huddled behind the wall for shelter from the rising wind coming from the sea) would have been overwhelmed by the sudden battering ram of force that had brushed aside the wall. But the greatest danger would not have been the force of the wind. Rather, the debris carried from shattered mobile homes and the shards of broken glass slicing through the air, at any other time of day, when people would have been up and about, would have caused many serious injuries. Those sheltering in the petrol station were doubly fortunate. Not only did this maelstrom of destruction miss them by a hundred metres or so, it could have caused the total destruction of the petrol station, its equipment and their cars. The time of year meant the caravan site was empty of people and the time of day meant no one was out and about where they would have been most vulnerable. By any measure this had been a near miss.

The tornado did not stop at the Youghal to Cork road. It had continued inland where the ground rises steeply to form a hill around which the town of Youghal is wrapped. On the lower and middle slopes of the hill is a medium density housing area. Following the route taken by the tornado, work in progress repairing roof damage could be observed. On close inspection this damage was most curious. The roof tiles had not been ripped from the roof as might happen in a normal gale. Instead, the tiles were partly stacked on the roof as if having been sucked underneath one another. The repairers had never come across anything quite like this before. To them this damage distinguished the winds of this storm from any other. The trail of roof and minor garden damage eventually stopped near the summit, since that was occupied by Youghal Golf Club and was covered in greens, fairways and open grassland. Beyond the summit and the golf club the damage was much less and the track petered out on the north facing slopes over the hill (Figure 1.2).

That Wednesday morning, the weather data being recorded at one of the Cork gas platforms in the Kinsale Area gas fields off the coast must have been the cause of some raised eyebrows and wry comments.

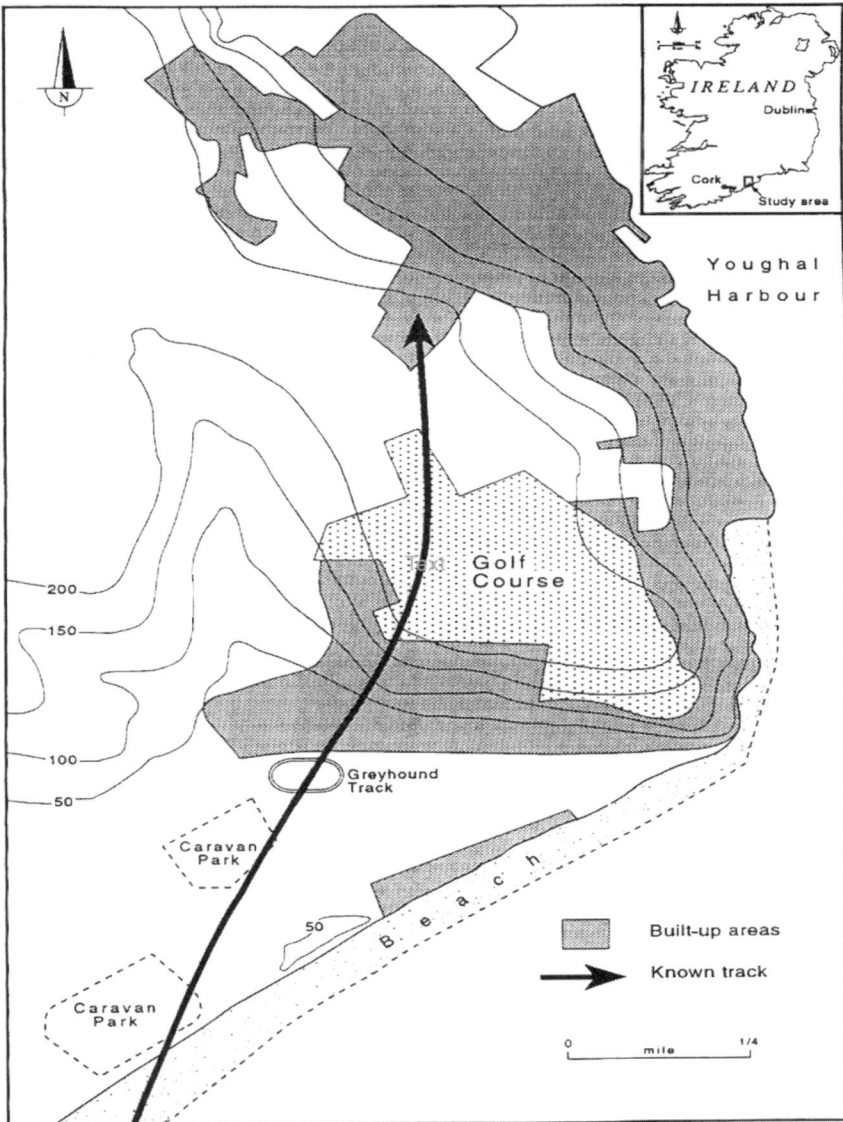

1.2 The track of the Youghal tornado.

By 7 a.m. the air temperature had reached twelve degrees Celsius. For a near-dawn temperature in February this was unusual, to say the least. However, the long-term records available for such offshore locations are minimal and any comparisons with the very different environments of the adjacent land areas are problematic. So, surprises were to be expected.

This must have been one such surprise. In terms of warmth, it was a most agreeable one for those on the drilling rig. Looking at the sky, an observer on the rig may just have been able to see in the early half-light a boiling sky of giant cumulonimbus clouds with shafts of torrential rain and the odd crash of thunder. Relatively warm moist air, even in winter, tends to produce such thundery weather. That was not so agreeable.

The wider weather conditions were recorded on the synoptic weather maps for that morning. On these were plotted all the weather observations and measurements from across the North Atlantic to mainland Europe, for given times. Those dawn hours showed a situation that occurred frequently. There was a depression (i.e. a low pressure system) off the north-west coast of Ireland, around which the surface winds were blowing in quite an usual way, drawing air in from the warmer south and colder north as it moved from the Atlantic towards Scotland. Every schoolchild who has studied weather in their geography or science classes would recognise the normality of this. In terms of the weather over Youghal, this resulted in winds from the south being drawn towards the centre of the low pressure. They were quite warm. However, what was interesting on this February morning was that hundreds of metres above the ground were winds that were stronger, colder and from a different direction. Between the lower and upper winds there was the potential for strong rotation to develop. This could then be stretched by the strong convection caused by the significantly warmer air being beneath very cold air. Thus a series of strong convective storms were generated in massive cumulonimbus clouds many kilometres across. These produced torrential rain and hail. But in one of them, a spinning column developed, induced by the wind arriving from different directions at considerable speed. This descended from the base of the cloud as it moved towards the coast. The column probably hit the sea surface before it reached the shoreline and would have appeared as a tall waterspout, mostly hidden from the shore by a curtain of torrential rain ahead of it and drawing up thousands of gallons of seawater which it sent spinning upwards before cascading away from the vortex. Then it hit the shore with destructive force and at that point became a tornado.

By the time of the visit to investigate the reports of a tornado, the local Gardaí were dismissive of the whole event. 'There have been others', they said. 'A couple of years ago something similar took off the roof from a building down on the quays'. Unfortunately, further enquiries yielded no information about that event, although a vague recollection seemed to be common. Indeed, from conversations with local residents there emerged strong evidence that such events were known to have occurred from time to time, but very infrequently. The dismissiveness of the Gardaí was of a different kind to the disclaimer that came later from professional scientists, who said that this was a 'freak'. But popular views tended to differ. This posed a challenge. Could they be right? The evidence from the recorded weather data and synoptic charts showed clearly that conditions had not been so unusual as to be unrepeatable. If so, a tornado such as that spawned by the conditions on that February morning could also have been produced on previous occasions in similar circumstances. More important than that, such events could occur again in the future and pose a hazard that was being completely overlooked.

Thus began a quest for a more complete picture of tornadoes in Ireland.

CHAPTER TWO

Encounters with Tornadoes

Nothing in the earth's atmosphere, perhaps nothing in all of nature, so uniquely combines the spectacle of terror and random violence against unsuspecting and innocent people as the tornado. Few other phenomena can form so quickly, vanish so suddenly, leave behind such misery, and yet still be seen as beautiful.[1]

Tornadoes are always unexpected. In Ireland they burst upon unsuspecting communities and people hardly know how to react. The reactions are usually deeply emotional and psychological. Often when the tornado is first seen in a dark, stormy sky, emerging from a storm cloud, the onlooker's mind and emotions have already been pummelled by thunder and lightning, torrential rain (possibly hail) and loud, driving winds.

The emotional response is often shock and fear, especially if the descending funnel is close by. There is the bewildering confusion that comes from sensing there is a threat and yet having no previous experience to deal with it. What should I do? Am I in danger? Should I run? If so, when? Where? Since there is little or no expectation of meeting a tornado in Ireland, few have given any thought about what they would do if it happened.

New Experiences: From fear to wonder

Close encounters with tornadoes in modern Ireland have mostly been when people have been indoors. Bad weather drives people to shelter and the rising, howling wind and other effects give no clues as to what may be approaching. Strong windstorms are common throughout Ireland and these past experiences dictate levels of anxiety, expectations of property damage and the awareness of personal risk. But where the tornado strikes, that past experience is overwhelmed and, as roofs are torn away, windows cave in and power goes out, there is almost no time to react. An intense chaos of sound and wind-driven debris occurs for perhaps thirty to forty seconds, maybe a little longer, depending on whether there is a direct hit. Then some sense of order and normal orientation returns. The shock and the fear that come with this maelstrom have had lasting effects on many in Ireland.

Psychological studies of children and adults have focused on post-traumatic stress disorder (PTSD) on the one hand and the development of phobias on the other.[2] Among children, natural disasters, particularly tornadoes, are as important a cause of PTSD as child abuse or loss of a parent. Fears associated with this disorder are also found in children who have not lived through a tornado, but have suffered from the effects of TV and films with tornado footage. It has been found that such fear and anxiety can be countered through empowerment (knowledge) and conversation (verbalisation) therapies. Among adults, even in the USA, it is not the severity of exposure to tornadoes that is linked to PTSD, but any exposure at all. There are a proportion of such cases following most tornado events, more so among women than men. This is likely to be of relevance in Ireland where it is often argued that the tornadoes are not as severe as elsewhere. If tornado size makes little difference, such effects will occur. PTSD affects emergency workers as well as the direct victims of such storms. Indeed, it has been found that tornadoes produce more PTSD symptoms among emergency workers than disasters such as when buildings explode.

Severe weather phobia is another effect of experiencing a tornado. The term 'severe weather phobia' was first introduced in 1996 as a result of studies of intense fear of severe thunderstorms and tornadoes. That fear leads to particular patterns of what appears to be precautionary behaviour. But normally such precautions are far from adequate and they still leave a lasting feeling of extreme vulnerability. Studies of these psychological conditions have been lacking in Ireland, although there is evidence from first-person interviews that these conditions do occur within the population here.

While fear and anxiety have been both immediate and long lasting for many people who have experienced tornadoes, it has not been so for everyone. Instead, for some there has been excitement and a massive adrenalin rush. The wow! factor kicks in and blots out every other response, at least briefly. Particularly from a relatively safe distance, eyewitnesses have spoken of being frozen to the spot in awe, wonder and considerable disbelief. The disbelief has come first, especially for a first experience in Ireland. 'It can't be ...', 'Surely that looks like a ...'. In an effort to persuade oneself, all other alternatives to a tornado race through the mind. Any look-a-like phenomenon will do. But as conviction dawns, awe and wonder often take over from the disbelief.

Are tornadoes a part of nature's grandeur? Few eyewitnesses can deny that as the slowly evolving, writhing funnels descend and grow to become tornadoes, often from spectacular parent clouds, they are a magnificent sight – at least from a distance. They fill eyewitnesses with awe as they begin to overwhelm the senses. Visually they have been described as beautiful in form as well, in their lighting effects and as they grow so improbably, but proportionately. Their texture and colour constantly change as they swirl beneath the parent storm cell with lighting effects often coming in from the side. Gradually their frightening power becomes more apparent and combines with their awesome beauty. Awe and wonder are not always to be found in calm, colourful, brilliant skies, but also in the power and supremacy of the storm. Nevertheless, some tornadoes are poorly sculptured, ragged and ugly. Others are made relatively invisible by curtains of rain. These show that awe may come from power without beauty.

As it approaches, the sound of the tornado blots out all others. All the normal sounds that make up everyday life, whether natural or the noise of human activity, are drowned out by the roar of the tornado grinding its way forward. Sound changes everyone's world and this unfamiliar one seems so out of place. It is a new sound of the power and energy of nature. Even funnels descending from their parent cell have been known to produce distinctive tornado sounds. This is because the powerful, swirling winds, whether surging along the ground or in the air above it, involve a lot of shear and turbulence. It is this, together with the sound of debris slicing through the air, that gives tornadoes their distinctive roar. In the USA, recent research has shown that funnels can produce distinctive sounds up to ten or twenty minutes before they reach the ground surface. On the ground, tornadoes have been heard up to 6.5 km away, although for the smallest 0.8 to 1.6 km is more typical.³ In the latter case, sound is probably the last of the characteristic signs of a tornado that reaches most people nearby. In Ireland, noise has been a characteristic of night-time tornadoes that have passed close by homes. During the daytime many have spoken of a silence and calm that preceded the tornado and its roaring wind. The silent zone is associated with an updraft region immediately beneath the storm cell close to the tornado. This is short-lived, from seconds to a few minutes at most. It is eerie and it has often burnt itself into the memory of those who were close enough to experience it. Such a dense silence has often made the onset of the tornado's roar even more stark.

Today, a growing reaction to seeing a tornado is to record it. The apparent compulsion to photograph a tornado whenever possible is understandable because it is a unique lifetime event. But it is also a way of producing convincing evidence to other people who would otherwise not believe a personal story. Many of the results have been stunning. Of course, some photographs are taken for scientific purposes and there is little doubt that research has been enhanced by having such data available for study. Stills and videos have captured a great deal of the detail of the tornado, its parent storm cell, vortex development, inflows and outflows at different levels and numerous other features that would otherwise not have been recorded. The use of these in the media has raised public awareness beyond the localities of the events to wider, sometimes national, audiences. This type of evidence tends to increase the sense of drama and threat compared to a verbal report. The images are much less used to emphasise the wonder and awe experienced by eyewitnesses. Indeed, in the media the images are frequently reinterpreted for effect by the headline and caption writers. There is a downside to all this, however. The apparent convincing proof of a tornado provided by a photograph has led to hoaxers using computer software to produce fake images intended to mislead. This has happened in Ireland. Therefore, no photographs can be taken at face value, but needs careful verification.

The effects of a tornado spread far beyond the physical dimensions of the vortex. Many more people see them than are actually caught up in their rampage across the countryside. From wherever a tornado can be seen it commands a visual space. Sometimes this covers a considerable distance. Within this space people may be awed or psychologically traumatised (or both). Such effects occur with ever-increasing intensity with increasing closeness to the tornado. In the Midwest of the USA this visual space is huge because the landscape is often flat and open. There, where skies are so big, the boiling clouds from which the tornado drops, the wedge or rope-like vortex curving to the ground and the swirling debris cloud at the surface, are all clearly visible. But in Ireland and many other parts of Europe the view is not as simple as this. What we might see and experience is often very partial. It is limited by nearby hills and mountains as well as other conditions. This adds considerably to the uncertainties surrounding the event and modifies the responses people make. Often only part of the storm can be seen and that may be very unclear. So reports of tornadoes may be very fragmentary. Given that they are unexpected, it is often difficult to be sure a tornado has occurred at all.

Was it a Tornado?

On 17 July 2007, a young girl, Lucie Shields, was standing at the side of the GAA pitch in Coalisland, County Tyrone. From the terracing she saw in the distance, towards Lough Neagh, the shape of a wide tornado stretching from the clouds to the ground somewhere beyond the nearby hills, where thunderstorms were rumbling.[4] Amid the excitement she had the presence of mind to take a photo. This was to be quickly acclaimed by the media as a tornado. Meanwhile, 35 km away at Maghera, Martin McKenna, an amateur astronomer and weather enthusiast, was watching the same thunderstorms and recorded that 'a very dark funnel suddenly dropped out from the base of a gigantic anvil'. He also photographed it, while his friend, Conor McDonald, captured it on video. All this excitement was packed within a short period of fifteen minutes or so, as they watched from very different locations. From each vantage point it was thought that the tornado was only a few miles away at most. The view from each location appeared to show that the vortex reached the ground surface. Subsequent field investigations based on line of sight

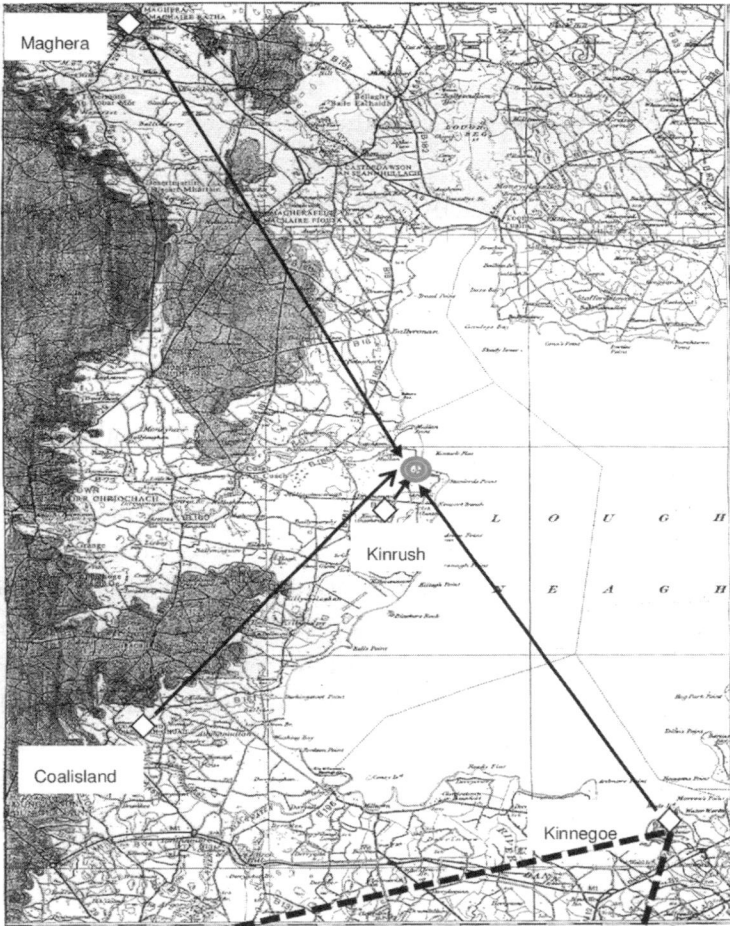

2.1 Lines of sight of various photographers to the Kinrush funnel cloud.
(Photo from Kinnegoe by Paddy Prunty)

mapping showed the funnel was as much as 25 km from Maghera and 15 km from Coalisland (Figure 2.1). Both distances were quite different from the initial estimates made by the observers and those who viewed the photographs subsequently. But of even greater significance was that when the locality of the funnel was confirmed and further eyewitnesses and film obtained, it was clear that the vortex did not reach the ground surface at all. The perception of a cloud to ground vortex had been created by the terrain falling away beyond the summit line of hills that were closer to the observers than the vortex itself. But at the surface beneath the funnel the wind conditions were quite normal and unaffected by the vortex several hundred feet above. At Kinrush, within the relative safety of this space, the O'Hegan family photographed the developing funnel almost above them and no more than a kilometre away. Their curiosity and amazement at what was unfolding was tinged with an element of anxiety in case it should descend and come their way. Their recorded snippets of conversation suggest they had one eye on possible escape routes, but such would be a last-minute response of dire necessity. Even further away, on the south-eastern shores of Lough Neagh, the staff of the Lough Neagh Discovery Centre watched and photographed the event from the comfort of the jetty at Kinnegoe on Oxford Island. They were equally captivated by the entire event. These sightings are shown in Figure 2.1.

Towards a New Awareness

Interviews with people in Ireland who have had an encounter with a tornado show that they have become changed people, often deeply changed. Many have recognised this and have spoken of having a deeper respect for the forces of nature; others of an increased sensitivity to changing weather conditions; while yet others of a heightened anxiety and tension when storms are expected. Few appear to have PTSD, although this may well be due to not having sought assistance, since elsewhere this has become a known common effect of encountering a tornado.

Even those who study tornadoes and arrive at the site some time afterwards are never quite the same again, despite not having had a first-hand encounter. A site investigation involves considerable attention to detailed damage and other effects. Recording mature, decades old trees with sturdy trunks that have been snapped in two some 3 m or so above the ground, as though they were dry twigs, can be a humbling experience. Sometimes they are twisted and torn as if a giant hand from the sky had grabbed them, casting the upper parts aside across the fields, leaving the twisted, jagged remnants pointing up to a now blue sky. From field to field a trail of immense chaotic destruction leaves you wondering at the real power of nature and the smallness of humankind.

For the student and researcher of tornadoes there is the challenge of exploring the full reality of the phenomenon. Tornadoes are not just the product of physical processes in our environment. Even if we knew all the physical ingredients that go into producing a tornado and its behaviour it would still not be fully defined and understood. For the student and researcher needs to engage with the power, beauty and awfulness of the tornado as experienced by the communities it affects, if there is to be any hope of fully understanding it and the threat it poses.

But social and cultural attitudes of Irish communities are a curious aspect of tornado study. The way people react to a tornado will be affected by many cultural practices of their own, and not merely by the nature of the tornado and its effects. The first thing encountered during this research was that there had been a big silence about tornadoes over decades and even centuries. It was as if there was a conspiracy of denial that such events occurred at all. The older generation in Ireland, particularly in the countryside, have frequently mentioned how they were afraid to speak, even in their own local community. Their reluctance was related to fears that their social credibility would be undermined, and they feared being thought odd. This phenomenon was particularly apparent in remoter areas where the population was sparse and a vortex might be witnessed by a single person. So their knowledge and experience were not passed on to others. In more modern times, probably the most frequent acknowledgement of the existence of tornadoes has been to refer to them dismissively as freaks. Happily this silence has now been broken. But it has led to a knowledge deficit about tornadoes in Ireland. There has been little generational inheritance of knowledge and understanding to guide us and shape our perceptions of the place of the tornado in the Irish experience of weather and climate.

This is quite unlike our knowledge of the seasons and of individual events such as thunderstorms, rainstorms, windstorms and other aspects of severe weather. These produce a popular understanding and wisdom, some of which has passed into our formal education system and has long been part of our understanding of our world. But not so for tornadoes.

However, the big silence is now being replaced by the big noise. The new willingness not only to speak of tornadoes, but also to report suspected tornado events, poses new challenges. Perhaps this is part of a necessary psychological response. It certainly indicates a greater confidence and trust by eyewitnesses that they will be believed. Trust is an important aspect of the reporting environment. Until the late 1990s this sometimes consisted of local media and only very occasionally national media. But reports also went to national scientific and research institutions such as Met Éireann and University College Cork (the latter being where this research has been conducted). The mobile phone revolution increased the number of people carrying camera technology which, with the video phone, expanded the recording possibilities immeasurably. Where words had been inadequate in trying to convince others of one having seen a tornado, the photographic record was seen to be definitive. All this resulted in more reports of tornadoes than ever before. But what was being reported? With a new confidence that tornadoes really did occur in Ireland, highly localised wind damage is increasingly reported as possible tornado damage, while scud beneath cloud bases is frequently photographed and reported as funnels or tornadoes. So the modern encounter has a strong popular technological element to it. But the encounter will always ultimately be personal and even emotional. Whether this will encourage a recovery of awe and wonder and encourage a greater awareness of the sky, its mood and its meanings, remains to be seen. If it does so, we will become a humbler generation. As Einstein said, 'What I see in Nature is a magnificent structure that we can comprehend only very imperfectly, and that must fill a thinking person with a feeling of humility'.[5]

CHAPTER THREE

The Early History of Irish Tornadoes

Climate is the full variety of past weather and the sum of our expectations of present weather.

One of the 'new' facts about the Irish climate established in recent years is that tornadoes are a normal, if infrequent, occurrence. They should, therefore, be included in any full climatic profile of the country. Indeed, in some years there are more tornadoes than snowstorms or cloudless days, while most years have more tornadoes than those larger destructive storms that occasionally sweep in from the Atlantic with winds of force 10 or greater. A history of tornadoes in Ireland must, therefore, look to the past with a reasonable expectation of finding that they occurred in earlier centuries as well. If they are part of our present, then it is reasonable to begin with an expectation that they have been part of our past.

The Earliest Records

How would a group of people hiding from a storm in medieval Ireland have described it if, in the terrifying crescendo of a severe thunderstorm, they saw what today we would call a tornado? Or, if a large, dark, turbulent cloud was observed slowly developing a narrow extension towards the ground, what words and phrases could have been used to tell of it? The technical vocabulary of clouds and weather mostly belongs to the nineteenth century onwards. Before then, what appeared in the sky was described in other ways. One of these has long been familiar to us since our childhood days when we traced the outline of animals, faces and other objects in the clouds as they passed overhead. For many centuries poets and writers have used this to advantage and have described the sky with vivid images derived from other experiences.

Thus, Dublin-born Jonathan Swift (1667–1745), author of *Gulliver's Travels* (1726), was able to describe a sky as follows: 'there is a large cloud near the horizon in the form of a bear, another at the zenith with the head of an ass, a third to the westward with claws like a dragon'.[1] Where was such an imagination cultivated? Making pictures out of the clouds may well have filled many a daydream while he neglected his studies at school in Kilkenny and later as a student at Trinity College, Dublin. No doubt his imagination was enhanced greatly by his unlimited access to the very large personal library of Sir William Temple, when he moved to England in 1688. Thereafter Swift returned to Ireland, to a number of ecclesiastical positions and became the Dean of St Patrick's Cathedral in Dublin. So the skies he knew best, and which triggered his imagination most, were Irish skies.

Not so William Shakespeare. He used the common thought forms of his public and struck a chord with them in the way he used observations about the weather and the skies. He borrowed an English skyscape for Antony in *Antony and Cleopatra*. But that mattered little to his audience who would have been familiar with his reflection in *Antony and Cleopatra*, Act 4, Scene xiv: 'Sometime we see a cloud that's dragonish, A vapour sometime like a bear or lion, A towered citadel, a pendent rock …'. Such metaphors and similes abound in literature. In the absence of a standardised, clearly defined scientific terminology, these images have had an important role in recording past skies. Such images often captured significant details of what had actually been seen. They provided a means of defining characteristics such as shape, size, motion, colour, texture and density. In this way the metaphor was used to set a particular sky apart in a very distinctive way. As a result, such descriptions have the potential for us of being regarded as a direct observation of a complex natural event. This is significant because observations and descriptions of all kinds have been a foundational key to the development of science as we know it today. Observation has always preceded understanding.

We must bear all this in mind as we go backwards in time in search of tornadoes. It is to the oldest historical Irish documents that we must go for records that may be the earliest reports of tornadoes in Ireland. These documents are known as the *Irish Annals* and were a product of the early Irish church. After Christianity was brought to Ireland (AD 431), the monks used what are known as paschal tables to create a calendar so they could compute the date of Easter. The tables had a manuscript line for each year, and it was not long before the scribes succumbed to the temptation of recording a few notable events against each of the years. Then, after a few years, the additions grew to embrace natural phenomena, often including notable weather events. At a later period all these records were collected together and incorporated into monastic chronicles in which events of special interest to the community were briefly noted as they occurred. There were many of these and from time to time such chronicles were in their turn collected together and then copied. It is these collections that are available to us today.

We may wonder whether such records have any reliability for us today. This is an important question if they are to be used to reconstruct details of the medieval climate in Ireland. Many of the *Annals* have been examined carefully by scholars for their accuracy, both in terms of their content and dating. In so doing a careful distinction has been made between two types of early record. There was an early native tradition of imaginative stories that were used and interpreted to fit current needs. But the entries made in the *Annals* could not have been more different. These were brief, bald statements of fact. For some time they have been regarded as the most authentic sources for Irish history from the fifth century AD onwards.

In more recent times the accuracy of the dates has been established from verifiable details such as the records of lunar and solar eclipses. This does not mean that the *Annals* are without problems and complexities that challenge modern academic enquiry. Not all of these celestial events made it into the *Annals*. They appear to have been included if considered by the annalist to be a portent of the 'Last Days' foretold in the Bible's Book of Revelation.[2] In addition, each of the *Annals* has its own complex history of aggregation and abbreviation by the copyists. Nevertheless, careful examination has concluded that weather and cosmic phenomena reports in the *Annals* are largely local after about AD 550 and that they must have been recorded originally shortly after they happened.[3] They also lacked the political and religious biases to which other types of entry into the *Annals* were subject.

In addition to the Irish documents, there were documents that originated far away but tell of events that were reputed to have taken place in Ireland. One such source lies among the literature of the Vikings. Not only did they raid Ireland, but from AD 795 they sometimes over-wintered in the country. They finally settled more permanently during their second invasive wave from AD 914. Of particular interest is *Kongs Skuggsjo* (it also has a Latin title, *Speculum Regale*), which is an Old Norse book, written about AD 1250.[4] In this book there are some amazing stories, similar to the very imaginative folk history compiled in Ireland that has already been dismissed as not being of value in the quest for historical information. These stories are mostly about the Vikings and little to do with Ireland. However, there is one that overlaps with events recorded in the *Annals* that may well be very important historically. The overlap is particularly significant because the two sources appear to have quite different origins and, therefore, are independent of each other. In *Kongs Skuggsjo*, the Irish personal names and place names are phonetic renderings, in Norse, of spoken Irish rather than the Irish or Latin written form. As a result, Irish scholarship has regarded this source as having been drawn from oral information rather than from written sources such as the *Annals*. It is further thought that such oral sources originated in Ireland itself and were independent of the *Annals*.

Irish Medieval Tornado Metaphors

This diverse literature gives expression to the rich cultural traditions that have contributed to the sources from which we are able to draw our knowledge of Irish history. Within it is contained a series of images, or metaphors, that were used with some consistency to describe atmospheric phenomena associated with severe storms.

The metaphors most used to describe a tornado were dragons, ships and towers, although there may well have been others. These metaphors would have been used by eyewitnesses reporting their experience to other people. The latter would have included those who made the record that has been passed down to us. In addition to these witnessed events were those that were not directly observed, but that struck fear and anxiety into those who suffered them as they sheltered out of sight. These events left a trail of damage and destruction that was sometimes recorded. To the modern eye these display a pattern that distinctively suggests the passing of a tornado rather than any alternative storm phenomenon.

Thunderstorms with dragons

Although not all modern tornadoes in Ireland occur in association with thunderstorms, on many occasions the meteorological conditions in which they occur are very favourable for thunderstorm development. Current data show that only 30 per cent of the reported tornadoes or funnel clouds have been in thunderstorms, although this is probably an underestimate.[5] In the *Annals* a number of severe thunderstorms are recorded. In two of these cases there is evidence that the thunderstorm may have produced a tornado. One is recorded in the *Annals of Clonmacnoise* for AD 734 and the second in the *Annals of Ulster* AD 735.[6] So these may have been two separate events in successive years. The similarity of the two records is striking. But the recorded events are considered to be relatively local and originally recorded soon after they took place so there is good reason to regard them as different events. However, it matters little whether this was so, because a detailed chronology is not the object here, but rather that such events were recorded at all. The two records themselves are very similar. In the year AD 735, the *Annals of Ulster* records that 'a huge dragon was seen in the end of autumn, with great thunder after it', while for the preceding year, the *Annals of Clonmacnoise* states 'there was a dragon both huge and ugly to behold this harvest seen and a great thunder heard after him in the firmament'. The noise associated with the dragon is clearly described as thunder rather than the roaring of the dragon, so there is little reason to consider that it was otherwise.

Interpreting dragons in ancient literature is fraught with obvious difficulties. They are often the stuff of fairy tales and consequently it is tempting to dismiss everything that mentions dragons without a second thought. But a serious interpretation of early medieval phenomena requires a careful examination of the evidence rather than dismissing what may be important information. It is striking that there are very few references to dragons in the *Annals*. Even in other types of early medieval Irish literature, monsters in general, and dragons in particular, do not appear frequently, and are not central to Irish mythology.[7] Indeed, they have little place at all, although we can find them today as symbols on a few remaining Celtic stone crosses.[8] The references to them in the *Annals* contain no detailed descriptions of these creatures. But it is not difficult to understand how a sinuous tornado funnel weaving a ponderous track across the countryside, spreading destruction, should be recorded as a dragon with its heavy body, huge outspread wings and long tail (Figure 3.1). Tornado funnels are not always small. The references to huge dragons suggest considerable dimensions. It is particularly worth noting that the record of dragons in the *Annals* is not only very limited, but it noticeably lacks the embellishments, battles and heroic deeds, as well as the social and cultural role of dragon stories, that are found in the record of dragon events in other countries, such as neighbouring Britain. It is from those latter sources, rather than the *Annals*, that we in Ireland have gained many of our images about dragons. In Britain, the role of the dragon in folklore was much stronger historically and the events surrounding dragon stories lack any association with particular weather types.[9] Even in a strongly dragon-rich culture such as China's, there is significant evidence that the dragon metaphor has been used for severe weather, and tornadoes in particular.[10]

A few years later another dragon event is reported in the *Annals of Clonmacnoise*, this time for AD 742. The record states boldly, 'there was dragons seen in the skyes', without any reference to weather conditions. It may well have been that, using the same format of the earlier record, this report is simply recording a series of tornado events. The plural form is used, so there may well have been more than one of these events during the year. Alternatively, more than one tornado may have been produced during a single storm, a situation that happens occasionally in Ireland today. The striking similarity of these brief accounts appears to bear out the argument that in the *Annals* convention was the most important determinant of what was written and of the form in which it

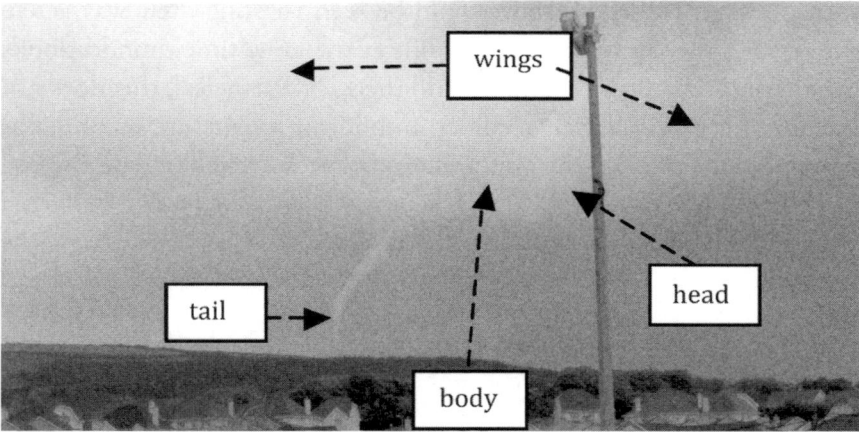

3.1 A dragon in the sky near Ballincollig, County Cork. (Photo by Tony Quane)

was recorded.[11] In this case the convention took the form of particular words and phrases – just as in the modern weather forecast!

Other possible alternative explanations for this imagery must be considered. The two most common interpretations for these ancient events that have been proposed by modern commentators are the aurora borealis (northern lights) or a comet. However, for both of these phenomena, good sightings require clear, cloud-free skies. The aurora borealis is seen in Ireland from time to time, although more so in the north than in the south. Such events are relatively infrequent, not least because of the general cloudiness so typical of Irish weather. The comet theory has been based largely on the image of the dragon's extended head, long body and tail when flying horizontally through the sky. But one of the most striking things when viewing a comet is the lack of apparent movement. To a person looking at a comet it seems impossible that something travelling so fast appears to be so static. It is only when carefully observed from night to night that its changing position becomes apparent. This seriously undermines the comet interpretation, as do the dates when matched to known major comets. In the *Annals*, the Irish dragon events are not embellished with additional details of colouring, light effects and other phenomena that can be used as evidence of a comet in the same way as they are elsewhere. Therefore, both the aurora and the comet interpretations for annalistic dragon events are much less satisfactory than identifying them as possible tornadoes. As tornadoes they would be seen twisting their way across the sky and the landscape below, while at the same time thunderclouds may have rumbled above and around them. Nevertheless, this does not exclude the possibility that comet and aurora events are recorded as dragons in the *Annals*. The interpretation of each event has to be assessed on its own merits.

Ships in the sky

From the middle of the eighth century AD the metaphor of the dragon in a thunderstorm appears to be replaced by a new and more informative metaphor: the ship in the sky. This new descriptive format may have been a result of an event occurring in quite a different locality to those from which reports had used the dragon metaphor. Perhaps this new locality was much closer to hand, even in Clonmacnoise itself. This is a very reasonable surmise because the time came when such an event did occur at Clonmacnoise, and this new image was used to describe it. The ship in the sky may be the most significant metaphor for a tornado used in medieval literature about Ireland. It is a very unusual image, but it appears both in *Kongs Skuggsjo* and in a number of the *Annals*. The fullest account of such an event is given in the former. In contrast, the *Annals* record them mostly as matter-of-fact statements, as if it would be obvious to any readers what these brief statements meant. The *Annals* contain five separate records of ships in the sky. The first three were within a few years of each other in the AD 740s, followed by a record of three ships in AD 763, and finally there is a similar record for about AD 950. Possible errors in the recorded chronology of these events, particularly in the 740s, leave doubts on their precise number and chronology.[12]

The first of these events is recorded in the *Annals of the Four Masters,* where the record for AD 743 includes '*Naves in aere vise sunt cum suis viris os ann Cluana maccunios*' (the translation being, 'ships with their crews were plainly seen in the sky this year at Clonmacnoise').[13] The other events of the AD 740s are in the *Annals of Clonmacnoise.* Despite the name of this manuscript, it cannot be assumed that all of these events took place in Clonmacnoise itself, but it is likely that they occurred within the Clonmacnoise region. It is noticeable how similar the wording is for these events. The phrase 'ships in the sky' is recorded with such brevity and economy of expression, with an almost stereotypical and formulaic vocabulary (lacking any attempt at explanation), that it is difficult to conclude other than that anyone reading the annotations would readily know what they meant. It is clear, therefore, that these features are presented very similarly to the way the *Annals* present the dragon events. So, again, convention appears to have been a strong influence in these records.

The Clonmacnoise tornado

The brevity of the descriptions in the *Annals* contrasts with the much fuller account of a similar event that is recorded, with the aid of the same medieval metaphor, by Meyer in his translation of *Kongs Skuggsjo.* It describes an event that occurred at Cloena (Clonmacnoise), and reads as follows:

> There is yet another thing that will seem most wonderful which happened in the city called Cloena. In that city is a church that is sacred to the memory of the holy man who is called Kiranus. And there it befell on a Sunday, when people were at church and hearing Mass, there came dropping from the air above an anchor,

as if it were cast from a ship, for there was a rope attached to it. And the fluke of the anchor got hooked in an arch at the church door, and all the people went out of the church and wondered, and looked upwards after the rope. They saw a ship float on the rope, and men in it. And next they saw a man leap overboard from the ship, and dive down towards the anchor, wanting to loosen it. His exertion seemed to them, by the movement of his hands and feet, like that of a man swimming in the sea. And when he came down to the anchor, he endeavoured to loose it. And then some men ran towards him and wanted to seize him. But in the church, to which the anchor was fastened, there is a bishop's chair. The bishop was by chance on the spot, and he forbade the men to hold that man, for he said that he would die if he were held in the water. And as soon as he was free he hastened his way up again to the ship; and as soon as he came up, they cut the rope, and then sailed on their way out of sight of the men. And the anchor has ever since lain as a witness of the event in that church.

All the tales in *Kongs Skuggsjo* are wonder tales. They are considered such, not because they are difficult to believe, but because they involve contact between the present world and another world.[14] But here the similarity between the Clonmacnoise story and all the other stories in *Kongs Skuggsjo* ends. A careful examination of the content shows that this record is more than likely rooted in actual historical events. For example, in medieval Ireland Clonmacnoise was a place of great learning, established by Ciaran in AD 545. It attracted many scholars and was described as one of the most illustrious schools in Europe. Events there would certainly have been widely witnessed and attested to with some authority, particularly in the *Annals*. Where an independent second source provides a similar record, as here, not only is the probable historicity of the event itself reinforced, but the additional detail it provides makes it possible to understand more fully the much briefer references in the *Annals*. *Kongs Skuggsjo* is not the only source of this expanded account. A briefer account of the same (or a similar) event, recorded in the Irish language, and attributed to an unknown author, has also been published.[15]

Some historians, puzzled by the record, but wishing to separate it from folk myths, have interpreted the account as a record of a vision. But in none of the sources are any of the events claimed to be visions, nor does there appear to be a history of visions in the Clonmacnoise community at that period. In any case, when visions did occur, they came to individuals rather than to large groups, and were usually indicative of spiritual authority or sainthood.[16] None of these conditions fit with the account. Indeed, an appeal to the physical and historical nature of the event is made at the end of the record, which clearly establishes that there was no intent to present the episode as a vision. It would seem inappropriate, therefore, for modern scholarship to impose such an interpretation on it, or to confuse it with a later event that may have been inserted a long time afterwards.[17] All this gives strong credence to the idea that the account refers to a physical event that recurred in a similar form from time to time and that had recognisable features.

The ship was not the first feature to be recognised. The onlookers first saw a rope and anchor falling from the sky. In fact it is clear that the rope and anchor were the most distinctive features of the episode. These descended from the cloud. Without a modern technical vocabulary this would be one of the most precise ways to report the development of a tornado. Indeed, a similar terminology is used to this today. Some tornadoes are so rope-like that they are identified as rope tornadoes. But more often, there is a rope stage in the tornado life cycle that marks the transition of a large tornado from its mature stage towards its dissipation. The Clonmacnoise tornado is likely to have been of the former type, a rope tornado, and its description predates by several centuries part of a vocabulary that is now widely used. It is interesting to note that many modern Irish tornadoes have narrow funnels and would appear very similar to what was observed at Clonmacnoise. The dissipation of the funnel also fits the description of the process that occurs when a tornado dies. It is described as being cut. Although many funnels withdraw into the base of their cumulonimbus parent cloud, funnels may also detach and break up in their final stages. The use of the anchor metaphor to describe the end portion of the rope emphasises that the funnel at the ground surface was much wider, and matches the debris zone of a tornado. At the surface, water, dust, soil and debris of all kinds are

3.2 A ship's hull with a descending anchor in the sky, being a funnel descending over County Antrim. (Photo by John McConnell)

levitated in the rotating wind field. It can be several times the diameter of the visible funnel and can be particularly enhanced when the visible funnel is narrow and rope-like.

Once a rope and anchor had been identified by the onlookers, it would have been very easy to see the cloud above as the underside of a ship. This would not stretch the imagination because many modern photographs of tornado-spawning clouds show a cumulonimbus cloud with a wall cloud protruding from its base, perhaps to a depth of 100 to 200 m. This gives the appearance of a ship's hull. In such photographs the tornado funnel extends from the wall cloud. In this way the image of a ship and its anchor is complete. Even without the wall cloud (they do not occur with every tornado), the cumulonimbus itself would be readily imagined as a ship, especially with the other characteristics mentioned (Figure 3.2).

But who, or what, were the crew that were observed at, or near, the cloud base? This has many possible interpretations. The possibilities range from humans to a variety of debris, but could also have been small individual cloud elements. With dramatic suddenness some of the large population around Clonmacnoise could have been caught up into the air, becoming merely a part of the lofted, swirling debris. Since the action of swimming implies sideways movement, it is possible that this detail indicates that vertically moving objects were rotating (a distinguishing feature of both tornadoes and funnel clouds). The anchor that remained could have been a larger item of debris caught up in the debris cloud that hit the monastery and remained there when the storm moved away.

In the Clonmacnoise event, the church was not destroyed. It probably provided some shelter and protection for those inside. This in itself gives an indication of possible dates when the tornado may have happened. It is highly probable that the storm predated the cathedral at Clonmacnoise (which was built in AD 904) as it is referred to only as a church. Nevertheless it would still have required a very strong structure to resist the wind speeds involved. Even the smallest tornadoes may have internal speeds of 64 to 112 km/h, but the occurrence of significant debris would indicate that the speeds were higher.[18] Many of the original churches in Ireland would have been flattened by such a storm since they were built of wattle and clay. Although by the eighth century many were being replaced by solid oak structures, or *dairthech*. Despite the fact that the influence of Clonmacnoise waxed and waned, it had been a monastic centre of considerable importance for a long time. As a result, it is likely that in the eighth century its church was a strong stone structure. If this was not so in its entirety, certainly in part. Such a structure would have been able to withstand the glancing blow of a tornado. Thus the account could match the ship-in-the-sky event of AD 743, which was specifically identified as having occurred in Clonmacnoise, or any of the other similar occurrences associated with Clonmacnoise. It is unlikely to have been before that date since the prevailing metaphor before this event was probably that of a dragon in a thunderstorm, as demonstrated earlier.

The Teltown tornadoes

A fourth event in the eighth century tornado record is of an occurrence at a place called Taillten (otherwise known as Tailtiu, and is the modern Teltown) in County Meath. This is recorded in another of the *Annals*, the *Book of Leinster*.[19] It states that in AD 763 three ships were seen in the air. This occurred when King Domnall mac Murchada was at the fair.

Tailtiu has been described as a city in manuscripts as early as AD 785.[20] However, it was likely to have been of considerable size and importance for a long time before that date. Its fair was of great antiquity and is considered to have been the most important fair in Ireland. It was held under the aegis of the King of Tara, probably the most powerful monarch in Ireland at that time. It lasted a full week, from the first Monday of August each year.[21] Just as Clonmacnoise was a major centre of religious significance where unusual events would be recorded, so Taillten was a place of similar significance for secular society, where a large number of witnesses would be able to verify the event. Little detail is given, apart from the fact that three ships are mentioned. This in itself is of considerable interest as it differs from the others in that those were single-ship events. The record may best be understood if it is assumed that the three ships occurred at the same time. It is well worth noting that wherever a tornado occurs multiple funnel events may occur as well. This is so in modern Ireland, as later chapters will show.

The fifth and final 'ship-in-the-sky' record is dated between AD 944 and AD 956. The year is not given but it occurred during the reign of Congalach. The source of this is more problematic. Originally it was contained in the *Book of Glendalough*, which is now lost to us. However, the record was one of many from that source that was entered into the *Book of Ballymote* and, as a group of the wonders of Ireland, became part of the *Irish Nennius*.[22] The text reads:

> Congalach, son of Maelmithig was at the fair of Taillten on a certain day, when he saw a ship in the air. One of the crew cast a dart at a salmon. The dart fell down in the presence of the gathering, and a man came out of the ship after it. When he seized its end from above, a man from below seized it from below upon which the man from above said: 'I am being drowned', said he. 'Let him go', said Congalach; and he is allowed to go up, and then he goes from them swimming.

This account echoes the Clonmacnoise event, but many of the details differ. This is reassuring because they are appropriately accurate for the locality where the event occurred. If this is a tornado funnel, it was seen as a dart thrown down at a salmon. In the tenth century, there were numerous artificial lakes around Taillten, constructed for fishing. So a descending funnel would have a high probability of striking water that was well stocked with salmon. The descent is described as a rapid one. This is often the case. Even wide tornado funnels may only take about a minute or so to descend. But witnesses observed a funnel that descended in a matter of seconds while they were observing a rotating cloud in County Cork on 18 August 2002. This could easily have been described as a spear-like thrust since it appeared to shoot out from the cloud base so quickly. The impression of a spear spinning through the air would have been enhanced by a narrow, rotating funnel, reaching the ground at an oblique angle (as occurred in County Cork). Again, the image of a man falling or swimming in the air would be consistent with debris or cloud fragments caught in the wider rotation wrapped around the inner spear-shaped cloud. This event is also more briefly recorded in the manuscript *Bishop Patrick's Poem*.[23] Patrick, Bishop of Dublin (AD 1074 to 1084), added a suggestive detail when he included the line, 'Who can hear this wonder and not praise the Lord of Thunder?'. This is indicative that the type of event was known to have traditional associations with thunder, as is often the case with tornadoes today.

Except for the Taillten event, the *Annals* record these events in the plural. In other words, more than a single event is suggested as having occurred in each of the years AD 743, 744, 749 and 763 (in the latter case three are mentioned specifically). If this detail is correct, then one of two possibilities must have occurred. There may have been a number of separate events during the year, probably spread over a number of months. Alternatively, there may have been at least one multiple event, where more than a single tornado occurred in a single storm system. Neither of these poses any problems in terms of the present climate of Ireland, so both must be considered possible. Even bearing in mind that the individual *Annals* would provide at most a regional rather than an all-Ireland record of events (and sometimes only a very local record), there is no reason, based on the grounds of probability, that more than one tornado would not have occurred. As will be seen in a later chapter, climatic data for the modern period shows that this holds good today.

Few commentators have engaged with the ships in the sky metaphor. When they have it has been to regard it as a wonder story and to dismiss any connection with actual historical events. Having made this assumption the similarities between accounts and the range of dates has led them to assume that there is a single basic version that has been repeated with some differences of detail. But this makes no sense of the record whatsoever. There is no purpose to be served by inserting the earlier references in the eighth century if there was only one later story. These earlier reports have no characteristic marks of having been added later, including, for example, the use of Latin rather than the Irish language.[24] It is a much simpler solution to regard these records as referring to something that actually happened on a number of occasions, and that these were recorded in a similar way over many years.

The tower of fire

Standing on the edge of the small County Westmeath town of Kilbeggan and looking in every direction, there is little to see. The terrain is flat, and the sky is big. On a good day far to the south the distant hills of Slieve Bloom might just be visible. For centuries the view would have been much the same. But on 30 April in AD 1054 it was suddenly very different. A remarkable sight was seen. Between a heavy storm cloud in the sky in the south and the ground beneath there was what appeared to be a huge tower of smoke. Onlookers would have assumed this came from a massive fire on the ground, the smoke stretching high into the air. Its slow progress towards Kilbeggan itself would have been a cause of considerable alarm. But this was no fire. It was a very large tornado whose description has been preserved for us in a number of the *Annals* and for many years has been one of the best known of the early European tornadoes. In the *Chronicum Scotorum,* which has its origins in a tenth-century Clonmacnoise manuscript that drew together a number of earlier documents, the account reads as follows:

> A tower of fire was seen at Ross-deala, on the Sunday of the festival of St George, during the space of five hours; black birds innumerable going into and out of it; and one large bird in the middle of it; and the little birds used to go under its wings when they went into the tower. They came out and lifted up, into the air,

3.3 A vortex appearing like a pillar of smoke, County Waterford. (Photo by Joe Cashin)

the greyhound which was in the middle of the town, and let it fall down again, so that it died immediately; and they lifted up three garments, and let them down again. The wood moreover, on which the birds perched fell under them, and the oak whereon the birds alighted was shaking, together with its roots in the ground.[25]

An analysis of this record, involving comparisons with modern tornado descriptions, has established that the rapidly rotating vapour funnel resembled the motion of smoke in a large fire; that the birds were either lofted debris or cloud segments rotating around the central funnel; and that the other lofted objects would only have behaved like that if caught in a tornado (another version of the event has the oak tree being carried off). It has been suggested that the stated duration of five hours is a copyist's error for the time of the event rather than its duration. Although this is a possibility, it cannot be taken conclusively.

This record also provides information about other aspects of the tornado. One is the tornado track. The tornado is first noticed over Ross-deala. That would be about as far away from Kilbeggan that it could have been, while still being seen from the town. At that stage it was well formed and had probably travelled some distance already. It was later in the middle of the town where it lifted a greyhound and other debris. The greyhound was probably a wolfhound, which were often called greyhounds and were prized hunting dogs kept in monasteries.

This is a direct distance of 3 km further northwards from Ross-deala. So the tornado's origin would have been some distance to the south-east, where there was no other significant settlement. From Kilbeggan, the tornado is likely to have continued northwards beyond the town. Being beyond the settlement, there was nothing of significance there that could be affected and noted by the monks.

Elsewhere in the *Annals* there are occasional references to other columns, towers or similar vertical features noted as having occurred in the sky, which may also have been tornadic events. For example, the fiery columns, noted in the *Chronicum Scotorum* for AD 944, may have been such events. As in the extended example above, a fire could well have been assumed to be the cause of the smoke-like vortex. No fire damage was mentioned in the record, unlike for other fire events.[26]

However, other interpretations of this event have been given. In particular, some of the details have been matched with detailed observations made of a supernova event recorded in China during July 1054, well known by historical astronomers.[27]

Tornado Events and Early Storm Damage Records

Overall, very few storms of any kind get a mention in the *Annals*. Even the most severe were only likely to be recorded if their destructiveness or the loss of human life made them exceptional. But tornadoes are dramatic events that are often destructive. The most severe of these would be just the type of event that the annalists would record if the impact was severe. In the first millennium, a number of probable tornadoes can be identified in the *Annals* from the clues the scribes provide about the nature and pattern of the storm damage they inflicted. These include events at Clonbroney (AD 783), Slane (AD 847) and Lough Hacket (AD 984).

The storm that was responsible for the first of these occurred during the night, when it would have been impossible to see much of what was happening. This storm destroyed the convent at Cluain-Bronaigh (Clonbroney), in County Longford. Cluain-Bronaigh was probably the first convent for nuns to be established in Ireland and was one of the most important at that time.[28] Its significance may well be the reason why its destruction was so widely reported and recorded. The geographical circumstances of the convent together with the nature of the destruction provide the evidence that the event, recorded in both the *Annals of Ulster* and the *Annals of the Four Masters*, could have been a tornado.

There are a number of reasons for believing that the convent was built of wood. Records show that the nuns had procured timber for building purposes in the eighth century and that the convent was burnt down in AD 1095, when it was probably still a timber building. In addition, the commitment of these nuns to poverty is consistent with an inexpensive timber structure. A wooden structure would have been much more vulnerable than a stone building to the destructive power of a tornado, although it would also have been vulnerable to the most severe gales as well. Clearly, the building material itself is insufficient to establish a compelling argument for a tornado to have been responsible.

However, it would have been normal for the area immediately around the convent to have been extensively settled, especially as the convent itself only owned a small area of land in accord with its vows of poverty. The convent was situated on a small hill, as can be seen today where the overgrown ruins of a later church can be viewed on the site of the original convent. But some hundreds of metres away to the north and also to the south, there are separate streams that would have provided the water upon which the poor and needy depended, and around which they probably clustered in their huts.

The account in the *Annals of Ulster* attributes the destruction of the monastery specifically to the wind rather than other causes. It reads, 'Terrible lightning during the entire night of Saturday, and thunder, on the fourth of the Nones of August; and a great and mighty wind destroyed the monastery of Cluain-Bronaigh.'

The destruction caused by the storm appears to have been over an extremely limited area. Had this been an extreme gale, such as those caused in Ireland by the remnants of summer hurricanes from time to time, the damage would have been much more widespread. But this damage pattern is very unlikely for an extreme gale event. Had there been widespread destruction and loss of life in the settled area around the convent, it is likely that such effects would have been noted, as was often done by the annalists. The failure to do so suggests the destruction was confined to a very narrow area, as would have been the case if a tornado had occurred.

A second probable tornado suggested by the destructive pattern of wind damage occurred at Slane, County Meath in AD 847. The case for this event is even stronger than at Clonbroney. It is recorded in the *Annals of the Four Masters* (and also in the *Chronicum Scotorum*), where it records 'the cross which was on the green of Slaine was raised up into the air; it was broken and divided so that part of its top reached Tailltin and Finnabhair-abha.'

Many early church sites in Ireland were marked by the erection of a freestanding cross made of wood or stone. The earliest of these would have been made of wood, but stone crosses became more popular in the ninth century. To both raise and break a stone cross would require an extremely powerful tornado and seems most unlikely. However, to lift and to carry a wooden cross would also be a substantial feat. The detail that the cross was broken would seem to favour a wooden cross (although a tornado snapped and lifted large grave headstones in Dunfanaghy graveyard during October 2002). Both the uplift and distance travelled in the Slane event require a mechanism that is only available naturally in a tornado. The feasibility of this is demonstrated in modern studies of long-distance transport by tornadoes. These have shown how a headstone was carried nearly 5 km in Minnesota, some turkeys nearly 16 km in Ohio and a cow also 16 km in Iowa (see Chapter 7). This reinforces the likelihood that this was something similar. The furthest fragment travelled from Slane to Tailltin (Teltown), a distance of 15 km.

Traditional Awareness

From early times the *Irish Annals* were largely a product of a monastic culture, so that Latin was always used. But from the start of the ninth century annalists were making extensive use of the vernacular.[29] With the arrival of the Vikings, Old Norse linguistic assimilations took place as well, no doubt reinforced by similar assimilations brought by the Anglo-Normans. Not surprisingly, a different vocabulary for describing tornadoes and other whirlwind phenomena developed. Sometimes these new words were used in translating the old records.

The first tornado event for which this new terminology was employed was quite early and can be found in the *Annals*. The *Annals of Clonmacnoise* uses the term 'whirlwinde' for an event at Logh Kynn, County Galway, in AD 984. The entry reads, 'The island of Logh Kynne, was by a great whirlwinde sonk on a sudaine, that there appeared but 30 feet thereof unsunkt'.[30]

The lough in question is now known as Lough Hacket. Today, there still a small island on the lough, about 50 m wide. The lough itself is only about half a square kilometre in area. The whirlwind is reported as an unusual event, particularly as it resulted in most of the island being 'sonk' with water carried by a whirlwind. A tornado over water is a waterspout, and this appears to have been what occurred. From the topographic detail of the island an estimate of the size of the waterspout can be made. The large, broad waterspout crossed the island from the lough and enveloped all but a small portion. Since the surrounding terrain is only slightly undulating and the water is quite shallow the distance referred to must be the horizontal distance: the portion of the island not engulfed by the event (about 10 m). Whatever the size of the island at the earlier date, the dimensions of the waterspout/tornado had to be considerable, since its circumference would have been more than 40 m. It is not surprising that the report describes the waterspout as a 'great whirlwinde'.

However, the word 'whirlwinde' may have been added retrospectively to this account when the text was translated. If so, it must have been a term in common use and considered to be a more accurate representation of the event in the record. It is known that from about AD 900 a number of Old Norse words had been introduced into the Irish language. The word 'whirlwind' is derived from the Old Norse *hvirfilvindr*, which appears in English-language documents from the mid-fourteenth century. Since the original text of the *Annals* has not survived it is impossible to know what it was translating. So it is uncertain whether it came from the settled Vikings or via the Anglo-Normans in subsequent translations of the originals.

The metaphors used in the previous accounts of tornadoes provided vivid pictures of what happened. But 'whirlwinde' was a more objective description of the wind's behaviour: it went round continuously in a circular motion. Although it could have been a textual modification, as suggested, there is no reason to suppose it could not have been in contemporary use. At some early stage, it would have been remarked upon that the cloud and the wind associated with the funnel were rotating. This could well have been in the late tenth century AD when the event was reported. It is likely that this must have occurred early on for the Old Norse word to be adopted.

This was the striking description of the probable tornado in AD 1137 recorded in the *Annals of Clonmacnoise*. Then, the wind caused large-scale destruction and debris, some of it of considerable size, and 'whirled some of them into the seas'. It is the whirling description that suggests a tornado was likely responsible for the damage. Certainly the sight of rotating debris was impressive enough to lead to its being recorded.

The Irish language vocabulary

It is a feature of language that when something is described that is familiar and important to people, a variety of words and phrases develop to describe it in its different aspects. For example, compare the number of words that describe different wind conditions in Ireland to those that describe its sunshine. Any dictionary will show that words for wind are far more numerous. Not only is this the case with wind, but more particularly the whirlwind. Over the centuries, the Irish language developed a number of words to differentiate between these. This is something that would not have happened if they had not been part of the lived experience of the Irish people.

One informative source of this vocabulary is the diary of Humphrey O'Sullivan, *The Diary of an Irish Countryman, 1827–1835*.[31] Humphrey O'Sullivan was a teacher from County Kerry, but was employed in County Kilkenny for many years. In his diary he carefully recorded details of life in the countryside and town, among which were many comments and observations about the weather. He noted five different ways of referring to a whirlwind in the Irish language, namely *guairdeán gaoithe, sián gaoithe, sí gaoithe, gaorsta* and *eachan gaoithe*.

To these can be added other words and phrases that have been recorded. In 1904, Lane in his *English–Irish Dictionary* had four expressions for a whirlwind: *sidhe gaoithe, gaoith guaisRéin, guadhsann/ guadhRann* and *gaoith suadh*. Successive dictionaries published during the twentieth century gave other words and phrases, each of which had slightly different emphases. So, in 1922 McKenna added *sidhe adhsann*, while in 1959 Tomas de Bhaldraithe in his *English–Irish Dictionary*, supplied separate words for a tornado (*camanfa, tornádo*), a whirlwind (*camfheothan, iomghaoth* and *cuaifeach*,) and a waterspout (*sconna, cuaifeach uisce* and *maidhm bháistí*).[32]

It is not the intention here to explore either the derivation or the emphasis of meaning in each of these words and phrases. It is sufficient to marvel at the extent of the vocabulary that has grown up in the Irish language surrounding the phenomenon of a whirlwind and its varied forms, and that this has been recorded over many years. In the English language, the phrase that for centuries has been used to translate and represent these Irish words and phrases is the 'fairy wind' or 'fairy blast'.

The faery wind tradition

There has been a long and ancient tradition in Ireland that the smallest whirl of dust or whirling whisps of hay on a summer's day, to destructive vortices that have flattened homes and even carried people into the air, are wind fairies or faery blasts. In the Irish language these are the *sidhe gaoithe* (or *shee-geeha*). The traditions that have grown up surrounding these show how much whirlwind phenomena have been bedded into Irish culture for centuries.

The anglicised word 'fairy' is very commonly used for *sidhe* to describe this phenomenon. This usage has contributed to its relegation into obscurity in the climatic profiling of Ireland. But these fairies are not the cute little things of popular anglophone culture. The alternative form, faery, still used by many in Ireland, is used throughout this account because it provides an immediate reminder as to how different were the *sidhe*. The *sidhe gaoithe* could cause death, injury, harm or illness as they passed by and, as a result, they had a fearsome reputation in rural Ireland. But there are numerous strands to the *sidhe* related stories that have been passed down over the centuries.

Traditional folklore tells how the origin of this goes back to the Tuatha de Danaan, a supernatural, angelic-like race who were conquered by the Gaelic Sons of Mil from Spain and fled underground to live in barrows and cairns (called *sidhe*, also meaning mounds or thrusts). The whirlwind was thought to contain the presence of the Danaan, so the *sidhe gaoithe* were literally 'thrusts of wind', regarded as bands of faeries travelling from one mound to another. They were dreaded more than loved, so respect was paid to them to avert evil because they were known to be able to harm cattle and damage crops. This resulted in their being called names that would not upset them, such as 'wee folk', 'good people' and 'fair folk' (shortened to 'fairies'). They were avoided at all costs and their homes were left untouched as, over the centuries, the land was cleared, and field agriculture spread. Whirlwinds of all sizes and widely differing destructive potential have contributed to maintaining these traditions. They are frequently revived when modern whirlwinds occur, be they tornadoes or dust devils.

The most common of these are the small vortices associated with otherwise relatively calm sunny conditions, probably best described as dust devils. These were familiar enough to be the subject of a painting by Cork-born Daniel MacDonald (1821–53) *Sidhe Gaoithe/The Fairy Blast*.[33] But the painting shows a small vortex. Much more significant were the life-threatening events that were recorded over the centuries. Some of these were described in terms of *sidhe gaoithe* events. Thus, Conchúr Ó Síocháin recorded the testimony of a fisherman in battling the great storm of 26 April 1894 off the south coast: 'From one glance I gave towards the stern I saw that moment a startling bright 'sheegwee' coming across the water, heading for the boat.'[34]

But most of the records available to us have been translated from their original form. In so doing traditional Irish-language terms were dropped in favour of 'whirlwind' or 'tornado'. Nevertheless, the use of the traditional terminology remains in many parts of rural Ireland, even today. This has also resulted in reinforcing the lack of distinction that these words and phrases make between significantly different types of whirlwinds, each of which takes on different forms, arises from different processes and poses different threats.

CHAPTER FOUR

A Prisoner of Consensus Science

Consensus closes the mind to other ways of seeing the world.

T he nineteenth century was an exciting time for the scientific community in Ireland. Traditional understandings were being replaced with new ones and a breed of scientist was emerging from a new rigour in observation and measurement developed in the preceding centuries. In Ireland, the Royal Irish Academy (RIA) was established in 1785, to promote this new science. One of its early presidents, Rev. Humphrey Lloyd, was strongly influenced by the meteorological agenda being pursued in several other European countries, to discover what were regarded as the laws of the atmosphere: the regular features that were predictable on annual, seasonal and daily time-scales.

A Limited Science

Humphrey Lloyd's meteorological research came to a climax when, in early 1850, the Council of the RIA adopted his proposal for a countrywide project to achieve these goals in Ireland. A network of sites was equipped with a common set of meteorological instruments to measure and record a range of weather parameters at set times each day. The year chosen for this was 1851, although some of the stations were set up and recording by October 1850. It was assumed that in a single year all the important variations would occur. As a result, it was expected that 'the phenomena and laws of storms' would be elucidated. The conclusions were presented to the RIA in two parts in 1853.[1]

Storms and the emergence of modern science

During the twelve months or so of this project, a number of significant tornado events occurred in Ireland. These were particularly noteworthy because they occurred in or near major urban areas and could hardly have gone unnoticed by most of their population. On 4 September 1851 the first of these occurred near Cork in the middle of the day.[2] The accounts of a number of eyewitnesses were recorded in the minutes of the Cork Cuvieran Society. Then, only four weeks later, on 5 October, a tornado ripped through the centre of Limerick with widespread damage and one fatality.[3] *The Limerick Chronicle* was one of a number of newspapers to report that such a phenomenon was 'hitherto unknown in this country'. Obviously, it was not well informed of events in Cork! Just one month later a report of this tornado was presented to the RIA by Dr Griffin of Limerick.[4] This was a detailed examination of all the reported incidents and damage associated with the event. As impressive as this tornado was in the overall history of tornadoes in Ireland, hard on its heels was an even more impressive event. On 24 January 1852 was the most severe tornado currently in the Irish record. This occurred in Nenagh, County Tipperary. With such a concentration of high impact tornado events there is every expectation that a record and interpretation would be contained in the report of the year-long survey prepared by Humphrey Lloyd, had Lloyd considered such weather events to be significant. But he did not. These events do not even appear as footnotes to any of the observations or descriptions of the individual storms that produced the tornadoes.

This raises intriguing questions about why such remarkable meteorological events should be so ignored in a study that was intended to provide a definitive account of the 'phenomena of storms' in Ireland. It cannot be explained by arguing that Cork and Limerick were so remote from Dublin that the events were insignificant to scientists in Dublin who had no direct experience of them. The 1851 survey was so designed to make the results geographically comprehensive across the whole of Ireland. Indeed, even scientists in Dublin would have had some familiarity with these rare meteorological events because in the preceding years a number of tornadoes and other whirlwind phenomena had occurred in the Dublin region as well. The most significant of these had been merely one year earlier when, on 18 April 1850, just prior to the national survey, a tornado had left an extensive trail of damage through the heart of Dublin and had terrified the population. The event had been the object of a detailed and careful study by Lloyd himself, who had reported on it to the Academy.[5]

Another possible explanation is that already there was an assumption, at least among the urban, educated population, that these storms did not belong to the Irish climate and were, therefore, an aberration. This may have been reinforced by stories of such events occurring in a variety of exotic, particularly tropical, places. In the Limerick event, the captain of one of the ships in the port, who had seen tornadoes before in the tropics, was reported as having recognised the phenomenon for what it was and alerting other vessels in the port to its dangers. In the Dublin event, *Saunder's Newsletter* described the tornado as 'so tropical and violent'. The *Dublin Evening Mail* associated it with almost any geographical region other than Ireland. It reported, 'Its phenomena were those peculiar to the sudden snow gales of the Baltic, the fatal Mediterranean white squall, or the disastrous and too often unforeseen and unprovided for West Indian hurricane'.[6] Each of these was quite different to what had occurred in Dublin, but the attempts to identify the event reflected an underlying need to locate its proper habitat in another place. Members of the Academy appear to have identified with the view that it was a geographical misfit of no consequence to their search for an understanding of the Irish climate.

There is another important reason why the 1851 tornadoes were excluded from what was intended as a comprehensive study of the Irish climate. The justification for a survey that extended into every part of Ireland, and that required the considerable commitment of resources and people for its proper execution, had to go beyond the rather abstract goal of searching for the laws of the atmosphere. The Great Famine of 1846–51 had only just ended when the survey was launched. Starvation and disease had devastated the population of Ireland and would traumatise the country for years to come. In promoting the 1851 survey, Lloyd had argued that science had to be justified by its relevance. The great social issues of food production and health were paramount, so the connections between weather, food and health became the focus of attention.

As a result, not only were tornadoes not referenced in the 1851 study, but other storm-scale phenomena received only slight attention as well. Mesoscale weather events such as thunderstorms and hailstorms were footnoted at the observing stations, but thereafter regarded as incidental. Lloyd's analysis ignored them all and their contribution to the Irish climate was largely overlooked. Further work by Lloyd and others reinforced this approach. At first his view was that the laws of the atmosphere could be established from only short periods of observation, but this was to change. Only a few years later he began to emphasise the necessity of gathering weather data over a longer series of years. In so doing he committed meteorology and climatology to the perpetuation of a research agenda that embraced only those atmospheric systems that could be caught in the sparse network of daily observations that developed across Ireland. This largely filtered out mesoscale systems.

Irish tornadoes go missing

All this was in accord with the developing international practice, especially across Europe. The International Meteorological Society was formed in 1873. This was the precursor to the World Meteorological Organization (WMO). While the resulting international cooperation achieved many benefits, one of the unforeseen results was that the whole concept of climate became relatively detached from day-to-day weather experience. Instead, it became fragmented into selected weather parameters that could be measured independently of each other with the increasingly sophisticated meteorological instrumentation that became available. The effect of this on the fledgling sciences of meteorology and climatology was to place a primary emphasis on the measurement of individual parameters such as temperature, rainfall, wind and pressure in order to identify normal conditions, rather than on building up climatologies based on mesoscale storm types and weather event chronologies. However powerful and destructive were these individual events, they became side issues, subservient to the task of gaining the wider, more generalised understandings of weather and climate that would help to harness the climate. In Ireland, the emerging climatic literature of the late nineteenth and early twentieth centuries strongly reflected this, as published for some regional climatologies, as well as for Ireland as a whole.[7]

This process had little or no place for traditional understandings of the weather and climate within Ireland. The *sidhe gaoithe* phenomena became a cultural curiosity of rural areas. But in the growing urban centres different understandings of atmospheric events were introduced that were to have far-reaching consequences that would extend forward, right to the end of the twentieth century. So, tornadoes that burst on to the weather scene in the 1800s, when the RIA was taking a detailed interest in understanding the Irish climate, were reported and then largely ignored. Although these were reported and assessed, there was a striking failure to make any reference to the traditional *sidhe gaoithe*. Instead, the accounts presented to the scientific community and to the urbanised public demonstrated quite different geographical associations. The most common association was with the tropics. As a result, the *sidhe gaoithe* phenomena were largely written out of the Irish weather experience. Consequently, even today, modern education has little place for them and fails to interpret the *sidhe gaoithe* in terms of modern scientific understandings of the same phenomena. This is not only a loss but sets science back many decades because it is unable to effectively gather the data of such events from the relatively rich variety of historical sources.

The twentieth century brought many new starts to Ireland. Independence and self-determination brought with them political neutrality and relative isolation from the European economic and cultural arena. Meteorology and climatology were fertilised by new forecast models and analytical tools from international sources. The Irish Meteorological Service (now Met Éireann) became part of the Department of Transport, reflecting a strong orientation towards the needs of air traffic and international aviation.[8] But this focus still provided no incentive to give tornadoes a place in the study of the Irish atmosphere. One of the earliest attempts at a continent-wide survey of European tornadoes, published in 1982, showed Ireland to be without any records of tornadoes and concluded that the country was relatively free of them.[9] However, this study failed to access the scattered, partial record of tornadoes that did occur, that could be found in limited circulation media or specialist magazines.

At the same time the Irish public was being slowly educated about the world beyond its shores, where tornadoes were recognised. But in encountering them in this way they were presented as exotic and distant. More than an average interest was taken in the USA because blood ties with North America had become particularly strong. During the course of the century, accounts of tornadoes from different parts of America gave the impression that they were very large, numerous and regular. Accompanying photographs emphasised their large size and the devastation they caused, and this seemed to be part of their essential nature. Not realising that these images were no more typical of the tornado in America than floods would be of every rain event in Ireland, it was easy to conclude that the tornado was not a part of the Irish climate. This reinforced the invisibility of Irish tornado events from the previous centuries.

It is undeniably the case that few actual reports of tornadoes or other whirlwinds were recorded in Ireland during most of the twentieth century. For many years there had been strikingly more tornadoes reported in the neighbouring states of Britain and France, and even in those countries the data suggest that tornadoes were under-reported.[10] However, on this basis alone it would be reasonable to expect more in Ireland. So why have so few been recorded here? There are a number of possible explanations for this.

The first of these is that many of the tornadoes that may have occurred were not seen because of the emptiness of the Irish countryside. At the end of the twentieth century the population of Ireland was still significantly less than it was in 1850. The population of rural Ireland continuously declined through the century producing a largely empty landscape where tornadoes would have done little or no damage that would impact people. The probability that a tornado would miss population clusters was high, so fatalities and other effects would be low.

Second, on occasions when tornadoes struck inhabited areas, they hit structures that were strongly built and would be largely wind resistant. Gale force winds and severe storms are very frequent in Ireland, and building design has tried to allow for these frequent severe conditions. Sturdy oak beams were used for important buildings from early times, soon to be replaced by stone. By the nineteenth and twentieth centuries, such storm-resistant building styles were widespread. The damage inflicted by severe winds was much less than if other materials had been used. As a result, reports of storms in Ireland that would have produced severe damage elsewhere (as in parts of the USA), would have been far fewer.

A third reason was the relative lack of economic impact. Tornadoes become noteworthy in a country when they cause serious financial losses. For example, the destruction of high value commercial crops can be a serious blow to families and communities whose livelihoods are affected. In this way a storm that is responsible for the damage gets reported. But in Ireland, where the principal crop has been grass, there are few areas that have been historically vulnerable in this way, with the possible exception of the south-east and east of the country.[11] But during the past century, Ireland's main economic crops have been grass and peat (with localised exceptions), on which a tornado leaves few traces. Disruption to the rural infrastructure was also of little significance as that tended to be both local and relatively basic in nature.

These demographic, cultural and economic factors have persisted and were demonstrated in the case of the County Kerry tornado of 6 June 1998. This was a weak tornado which occurred in a remote landscape of peat lowlands and forested hills. Subsequent research established that none of the scattered local population observed it as they sheltered in their homes from the stormy weather. But it was inspected closely from a passing helicopter.[12] As a result it was not reported in any public media and no awareness of the event developed until an investigation was initiated (see Chapter 5).

Thus, many reasons can be given for the invisibility of the Irish experience of tornadoes. But it can be argued that the most important factor of all has been the role of a dominant scientific paradigm that had no place for it. This became established in the nineteenth century, and it shaped the direction of climatic research from that point onwards. We have seen that in nineteenth-century Ireland, climatological work was dominated by a particular, limited agenda. During the twentieth century, new paradigms emerged from the old. From the end of the twentieth century the prevailing scientific paradigm in which the study of the earth's climates has largely been framed has been that of climate change. This is not to deny the importance of the growing attention paid to climate impact studies that have focused upon climatic extremes and hazards. But in recent years most of these have been related to the greater global questions of climate change. In Ireland, the focus of climatic research has continued to reflect this international research agenda and its priorities.

Tornadoes as freaks

One result of all this has been that until recent years in Ireland even large tornadoes that have occurred received little or no scientific attention. This has been compounded by a vocabulary of denial, using terms such as 'freak' and 'abnormal', that have been used by the press and other media. Such language has certainly helped to shape popular thinking. Typically, press reports have repeated this practice, in phrases such as 'freak storm', 'freak weather', 'freak whirlwind' or 'freak tornado', as well as in phrases with similar implications such as 'believe it or not' or 'miniature typhoon'.[13]

This was also a characteristic tendency of some scientific or other expert commentary, especially in the late 1990s, as commentaries from sources perceived to be scientifically authoritative became more frequent. Even in 2005 a Met Éireann spokesperson referred to a tornado report in County Mayo as a 'freak accident'.[14]

In all these cases there appears to have been a lack of a reference framework in which to place these events and to help understand them. Interestingly, this was not so much the case in traditional rural society, which seems to have been much more comfortable with these events. So, in the case of the Castlefin tornado in Donegal in March 1995, the tornado was described in the newspaper the *Finn Valley Voice* as 'a giant shebeen whirlwind' and 'the work of the fairies', drawing on the understanding that had developed over centuries referred to earlier in this text.[15] Although no official scientific comment was forthcoming on that occasion, a month earlier the tornado in Youghal, County Cork was described as a possible squall by Met Éireann, using the frame of reference that was familiar to trained meteorologists.

It is worth considering what is being said when an event is described as a 'freak'. A freak is defined as something capricious, or an unaccountable or unforeseeable event.[16] It will be immediately apparent that this is a particularly inappropriate word to be included in a scientific comment about the tornado in Ireland. It is only unaccountable in the sense that we have an incomplete understanding of the way the atmosphere works, especially at the scale of individual storm cells and storm systems. When it does occur it is a product of particular atmospheric environments in which a range of specific processes are at work. To dismiss such events as freaks is to discount them as being less than integral to our meteorological experience in Ireland. For many years the scientific community has contributed significantly to the perpetuation of this view, as demonstrated in Irish-based textbooks, scientific training and formal commentary.

The national media have reflected this view, so that reports of tornadoes have been confined largely to the local press, if reported at all. For example, the Wexford tornado of 7 January 1998 caused damage to industrial, commercial and public structures, although some of it was of a minor nature. The damage did ensure local press coverage, although it was ignored nationally. The tornado's track was about 8 km and was a significant tornado by Irish standards. But there was no national awareness of the event. The previous year a more remarkable event was a totally unreported (at the time) tornado on the outskirts of Cork city. This was a small tornado known to have been observed and eventually reported by three people who did not recognise what they had witnessed and did not report it at the time as there was no significant damage or injury. Since it was close to a dense residential area, there were probably other people who observed it. Thus, it became one of the invisible tornadoes, of which there must be very many that have left little or no imprint on the landscape or the public consciousness.

A New Awareness

If our thinking about tornadoes has been in the vice-like grip of a paradigm that at best marginalises the phenomenon and at worst continues to deny it, the task of bringing about a change to our way of thinking appears to be almost impossible. However, scientific thinking may be revolutionised when exceptions to the prevailing scientific viewpoint increase in number until it is eventually shown to be overwhelmingly inadequate.[17] Experience in Ireland has shown that as well as by the accumulation of exceptions over time, the same effect may be achieved by means of a defining event that brings a major new meteorological experience to the attention of the public and to the scientific community. The graphic images of the modern media provide a mechanism for such a process that was largely absent in past decades. For those thus exposed via the media to the event, there is a shared experience. This shared experience then changes the accepted frame of reference for expected weather events.

The Ballysadare tornado as a landmark event

The tornado that occurred in County Sligo on Friday, 6 August 1999 may have been such an event. What was special about this was that it was captured on video by Keith Sleater of Knocknahur, as he was arriving home from work in the late afternoon. His home overlooked the Ballysadare River estuary and, as he stopped his car, he was amazed to see a fully formed tornado crossing farmland on the opposite side of the estuary. He had the presence of mind to reach for his new camcorder and shoot his very first footage! Due to the initiative of Met Éireann's duty forecaster, some of this footage was shown the next day on the national television channel in the evening prime time weather forecast. It was the first time a video of an Irish tornado event was presented to the nation.

A new public awareness of tornadoes became widespread as a result of this presentation. This was demonstrated when a multiple tornado outbreak occurred eleven days later. It became the best reported tornado event in the Irish record up to that date and for a considerable period beyond it. Over a 24-hour period on 17–18 August 1999, nine separate tornadoes and funnel clouds were reported in the west of Ireland over the counties of Cavan, Galway and Mayo. For the first time sufficient reports were received to establish that this was a multiple outbreak. The reports were made by people who demonstrated a confidence in what they had seen that had been absent in reports before that date. No longer were they apologetic in nature. Instead of belittling what they had seen by using misleading terms such as 'mini-tornado', or almost excusing their observation by calling it a freak, eyewitnesses spoke positively and confidently of seeing a tornado. They reported their experiences with a clear belief that what they had seen could well have been a tornado (although frequently they wanted confirmation of this).[18]

More substantial evidence for this change of view was gathered through a survey carried out at the end of that August.[19] Those data showed that 80 per cent of 140 people surveyed within 10 km of where the Ballysadare tornado occurred had their expectations about other tornadoes occurring in Ireland changed. Also, nearly 70 per cent declared that they had expectations of other tornado events occurring in Ireland sometime in the future. However, some respondents' traditional views about tornadoes were quite unchanged, so a credibility gap remained. Of course, if the survey had been conducted elsewhere in Ireland the response may not have been so strong. It can be expected that continuing exposure to reports of these events, or to the events themselves, will be necessary to achieve widespread acceptance of their normality.[20]

In the footsteps of Beaufort

A major tool for studying tornadoes in Ireland was developed in the 1970s in the form of a scale to measure tornado intensities. This is an important means for differentiating between tornadoes, based on their severity. There is a clear link between tornado intensity and their destructive potential. The latter is of concern to people all over Ireland and determines how seriously such weather events should be taken. In Ireland tornado intensities are now assessed whenever possible. These have been based on the formal extension of the Beaufort scale of wind speed, familiar to all weather enthusiasts and professionals as well as being a standard scale for estimating wind speeds that appears in all climatology textbooks.

Although the Beaufort scale is used globally, in a sense it has its origin in Ireland. Francis Beaufort was born in 1774 in Navan, County Meath. But he went to sea at the age of fourteen and by 1790 he had joined the British Royal Navy. Despite being largely self-educated, he developed the first version of his scale for wind force in 1806 when in command of HMS *Woolwich*, undertaking a hydrographic survey off

South America. This scale correlated wind strength with the amount of sail a full-rigged ship would carry. It was fully adopted by the Royal Navy in 1838. However, it later went through a series of modifications to accommodate conditions on both sea and land.

Beaufort's scale of wind force was adopted for use with tornadoes in the 1970s. Then, the Tornado Research Organisation (TORRO) introduced a scale which used Beaufort force 8 (a gale) as its zero value for tornado intensities and applied the modern Beaufort wind speed formula to designate categories between 0 and 10, each being defined as a range of wind speeds, as summarised in Table 4.1. This tornado intensity scale (the T-Scale) was used from 1972 after a period of testing and the details were published in 1976.[21] This has been used as a basis of tornado reporting and classification for Ireland since systematic research into the occurrence of tornadoes in Ireland began during the 1990s at University College Cork. This was not merely because of the Irish connection with Beaufort, as attractive as that reason would be for research in Ireland. Rather, in comparison with the USA's Fujita Scale (or F-Scale, which was also developed in the 1970s but was changed to the EF-Scale in 2007), the T-Scale has the benefit of having more categories.[22] Early investigations of Irish tornadoes during the 1990s showed the T-Scale to be more useful since the most common tornadoes in Ireland were of small to medium intensity.

Table 4.1 The tornado intensity scale developed by TORRO.

T scale wind speeds							
T	metres /sec	mph		T	metres/ sec	mph	
T0	17–24	39–54	light	T6	73–83	161–86	moderately devastating
T1	25–32	55–72	mild	T7	84–95	187–212	strongly devastating
T2	33–41	73–92	moderate	T8	96–107	213–40	severely devastating
T3	42–51	93–114	strong	T9	108–20	241–69	intensely devastating
T4	52–61	115–36	severe	T10	121–34	270–99	super tornado
T5	62–72	137–60	intense				

Weather enthusiasts and technology

A greater awareness of tornadoes and their place in the Irish climate has grown in the modern period. This is partly as a result of changes that have made it easier for more people to record tornadoes themselves. In Ireland, weather enthusiasts, both individually and in organisations and networks, have played a pivotal role in reporting and recording tornadoes. This is a notable continuity from the past and connects with the amateur tradition of the nineteenth century and earlier, before the professionalisation of meteorology and climatology occurred. In recent years the photographic record has expanded enormously. Now, digital photography, mobile phone cameras and easy access to the worldwide web to post and send photographs all make important contributions. The growth of a community of weather enthusiasts, both professional and amateur, prepared to chase storms (even in Ireland) to see how they develop and what they produce, has also played a part. Each of these factors have contributed to building a database of verified events that has been exceptionally significant.

As demonstrated in the case of the Ballysadare tornado, the media had a positive and constructive role in developing this awareness. However, it has also developed a language that has led to misunderstanding and confusion. The terms 'mini-tornado' and 'twister' are two good examples of this. These terms have been used to mean quite different things in the reporting of events across the country.[23] Thus, a report of a tornado, from whatever source, is quite different from a verified occurrence. Verification has become an even more critical process than it was in earlier days. This normally requires a thorough site investigation, involving damage assessment and mapping, eyewitness evidence and a meteorological assessment of the event.[24] Many reported events have been found wanting as a result. However, these conclusions have often not been available until sometime after the event and can rarely be communicated to the public. Popular perception of the frequency of tornadoes is, therefore, only shaped by the original, often dramatic, reporting.

Little of these data can be gathered or verified through the national network of state meteorological stations. In recent years there have been up to eighteen synoptic stations reporting to Met Éireann each hour. These have fixed locations and record conditions at those specific sites. The probability of a tornado going anywhere close to one of these sites is extremely low and the staff are not equipped to carry out the necessary local survey work away from their meteorological station. Of even greater significance now is the fact that most of the meteorological stations have been, or are being, automated. At such stations staff would not be available to carry out the necessary site investigation work.

Tornado forecasting

The ultimate application of our understanding of tornado-forming processes is to be able to forecast them in order to appropriately warn communities likely to be in their path. Pioneer work on forecasting tornadoes was carried out in the USA during the nineteenth century. But this was hindered by the imposition of a ban on using the word tornado, which lasted for four decades, for fear its use in a forecast would generate public panic. This practice lasted well into the twentieth century. However, on 25 March 1948 the first tornado forecast was issued for an airport base in Oklahoma. It was successful on two counts. Firstly, the tornado occurred and secondly, although it caused 6 million dollars' worth of damage, no lives were lost. As a result, the value of such forecasts became well established.

One of the first successful tornado forecasts for Ireland was made on Friday, 25 October 2002. This was one of the early experimental forecasts that were being issued by TORRO when conditions seemed appropriate. This was circulated within the TORRO community at 08.15 UTC (09.15 GMT).[25] It was in the form of a Severe Thunderstorm Watch which recognised a possible tornado threat within the watch area, for the period 09.00 UTC on the Friday to 05.00 UTC the following day. The forecast area covered central and southern regions of the country. A series of tornadoes occurred in Ireland that day. The first was a small tornado near Preban in County Wicklow. Later in the day a further weak event occurred at 12.20 UTC at Castlederg in County Tyrone. But the most severe (and most spectacular) event was in County Longford. At 12.30 UTC, during an intense thunderstorm over Derrynacross, near Ballinalee, a tornado was seen developing on the

4.1 The concrete base, supporting blocks and wire cables from which a mobile home was ripped away at Derrynacross.

north-western side of Corn Hill. It swept across the hill to its eastern side where it crossed farmland leaving a trail of destruction. As well as flattening trees, it destroyed a farm shed and ripped up a mobile home anchored by steel cables into a concrete base, lifting it across a lane into a nearby field (Figure 4.1). The tornado crossed several farms before hitting Derrynacross forest, snapping and levelling scores of trees in different directions as the vortex span into it. Among the eyewitnesses was Joseph Donaldson, whose own home suffered minor edge effects as he watched in disbelief the tornado approaching and then engulfing the land at the side of his house, before heading away with the mobile home that had been there.[26]

While the verification of local tornado reports by on-site investigations has been the major means of testing tornado forecasts, a second means is the storm chase method. This methodology has a very limited history in Ireland because of the complex, narrow winding road patterns that make for an unfavourable chase environment. In addition, poor secondary road surface conditions are a constant hazard. Nevertheless, the storm chase method has important potential for providing records of events that would otherwise be missed because

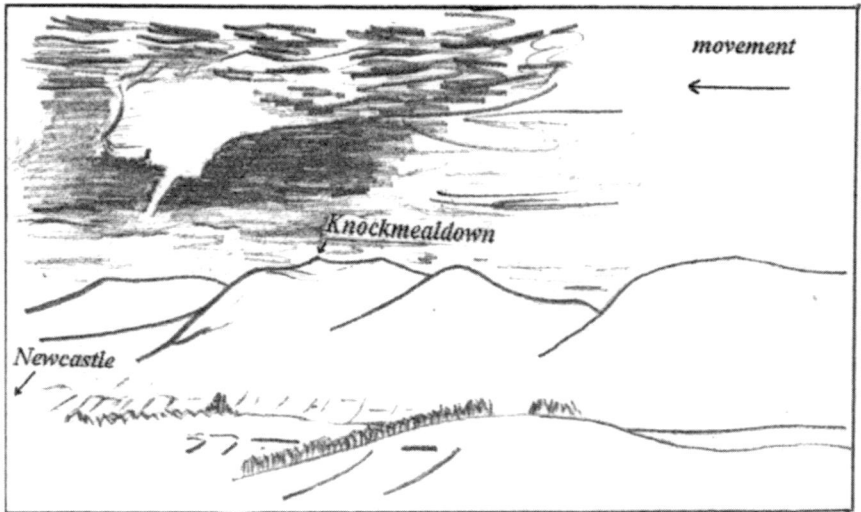

4.2 A field sketch of the funnel cloud made during a storm chase; view is from County Tipperary towards the south.

of the relatively empty countryside. Within the forecast experimental period, the very day before the events described above for 25 October 2002, such a methodology was used successfully to capture a funnel cloud in County Tipperary (Figure 4.2). Although there have been storm chases on earlier occasions, this is the first known published account of a storm chase in the country.[27]

It has become an important principle in establishing an accurate database that such events do not rest on a single report, even if it is a photograph, field sketch or similar record. A second independent report is the minimum required to verify a funnel cloud report. For a tornado, a site investigation is preferable to confirm the reliability of any other evidence.

After a few years of experimentation TORRO began to issue tornado forecasts publicly from 1 January 2006.[28] The forecasts have been for the area of Britain and Ireland. During the first months of the year the results were disappointing. Up to the end of March 2006 there were three dates that were identified as having a very slight tornado risk in Ireland, but there was no evidence of anything having occurred. However, on 31 March a warning of possible isolated tornadoes in

Ireland was accompanied by a map of the area considered to be at risk. The forecast time was between 14.50 UTC and 17.00 UTC. At 15.35 UTC a storm cell over County Cavan spawned two tornadoes close to Bailieborough. The larger of the two was described by local people as being ten times the width of the road while the second, which was very close to the first, was described as being about half that diameter. They produced a combined damage track of 7.2 km (4.5 miles) which at times was 275 m (300 yards) wide. Along this track, severe structural damage was caused, trees snapped in two and at one location a parked milk tanker was lifted and hit the side of its depot, damaging the wall and roof. Despite the hilly terrain and the poor weather this incident was observed by a number of people, but fortunately there were no injuries.[29] These results demonstrated that the forecast procedure had proved itself and its potential was clearly established.

For forecasts to be most effective, a significant lead-in time period is needed for warnings to be issued to vulnerable communities and activities, as well as time to allow appropriate actions to be initiated. Such timely warning can be more effective if there is precision about the geographical locations at risk. So, although major progress has been made, there is still a great deal to do in order to lengthen the lead-in time and narrow the forecast area. This can be achieved through a greater understanding of how tornadoes develop, and how this is related to the dynamics of parent storm cells and the behaviour of the weather systems that produce them. But, in addition to such scientific understanding, effective warning systems and communities that have the ability to respond to issued warnings are equally important.

New Dimensions

Thus, there have been a number of strands to the rediscovery and developing new awareness of the tornado in Ireland's climatic profile. This reawakening has been a gradual one. It has been a result of converging individual initiatives and more general trends. Among these have been four particularly significant ones. The first has been a willingness by individuals to communicate unusual experiences, initially to their community, but later to a wider audience. A second has been the application of a meaningful system of measurement of key characteristics that reflect impacts of significance, using a scale appropriate to Ireland's range of events. The third has been the development of technologies from the relatively simple and accessible that enable eyewitnesses to confirm what they see to others, to a variety of more complex research technologies. Finally is the experimentation and application of forecast methodologies appropriate to Ireland.

Each of these has contributed in different ways to the ongoing process of establishing the reality of the tornado as an Irish weather phenomenon.

CHAPTER FIVE

The Making of a Tornado

Can anyone understand the spreadings of the clouds? [1]

As the Irish Coast Guard (IRCG) helicopter rescue unit based at Shannon was returning from a mission over the MacGillycuddy's Reeks in County Kerry, in June 1998, it was flying below the cloud base and had a reasonably clear view of the Stack's Mountains just ahead. There was no heavy weather around them, but earlier there had been thunder. In fact on the ground, where farmers were trying to cut silage, the weather is remembered as having been quite violent. Now the helicopter was passing under what appeared to be a benign, but large, cumulonimbus cloud. As the crew of nine began to relax, the pilot called back to his winchman, 'here is another one of your things, dead ahead'.

This remark was a reference to what his colleague had seen only weeks before, as he drove through County Laois. Being interested in the weather, his attention was drawn on that occasion to the base of another cumulonimbus cloud that had become very dark. Eying it as he drove, he had noticed that the underside became turbulent, almost as if it was boiling. To his amazement, from this base a funnel-shaped cloud began to descend. He pulled in to the side of the road so he could watch the spectacle more closely. It lasted only a few minutes and to his trained aeronautical eye it only descended a couple of hundred feet, leaving it suspended well above the ground surface. Having told this to his colleagues on his return to Shannon, they now turned to him as they approached another funnel-like cloud extending from the cloud base

59

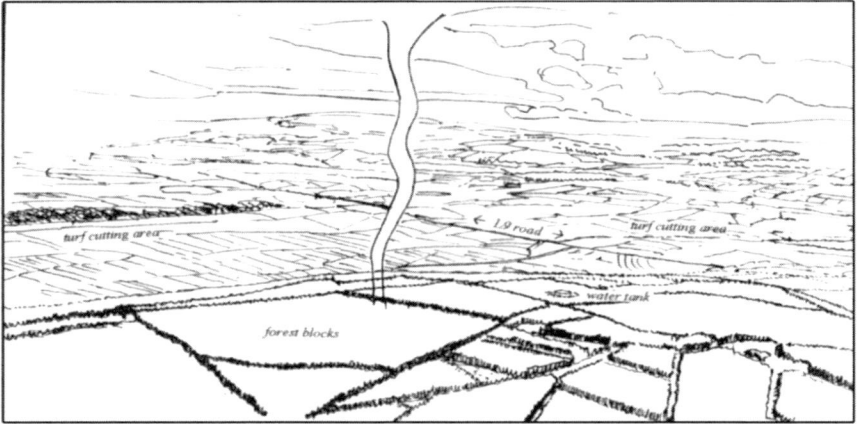

5.1 Sketch of the Stack Hills tornado based on the photo from the IRCG rescue helicopter.

right to the ground. In their minds, these things were not supposed to happen over Kerry!

What they saw was striking. It was a narrow cloud, almost tube-like in shape, and from the helicopter the diameter seemed only about 10 m, although it was probably considerably wider. It was also fading fast. As the helicopter approached, so it appeared to be thinning. Indeed, it seemed as if it was fragile, and the pilot was anxious not to get so close as to destroy it. Part of the training of the crew develops an instinct to record, and this was the reaction now. There was just one frame left in the camera they carried on rescue missions, so the helicopter circled the funnel cloud to allow this one, last shot to be taken (Figure 5.1). But fuel was running low. The helicopter crew could no longer stay and watch what happened, so they headed back to base.

Tornadoes are, by definition, violent. Few would describe them as being fragile. The helicopter crew perceived no threat; they were merely curious to see a phenomenon that had never crossed their path before. However, although this tornado was small, being in its final stages when it blocked the helicopter's path, it was still very dangerous. High speed rotating winds would have covered an area much wider than the visible funnel cloud itself. Had the helicopter entered that wider rotating zone, it would have been extremely vulnerable, and the consequences could have been serious. This was a near miss. The incident illustrates how varied tornadoes can be in Ireland. It also demonstrates the need to raise awareness of what they are and the potential threat they pose.

The Elements of a Tornado: Visible and invisible

When asked 'what is a tornado?', many people in Ireland have difficulty in separating it from other severe windstorm types, such as the hurricane. However, for most it is its distinctive appearance that is definitive. So, that is as good a starting point as any for exploring the tornado phenomenon.

What is seen in the air

A tornado is a violently rotating column of air. Its violence can be the cause of great destruction. Of course, there is more to it than violent rotation, but this is its basic essence. It is often mistakenly thought that the tornado is the characteristic column-like cloud. But this forms within a wider swirling mass of air. The cloud does no damage, whereas the wind does. There are other rotating structures in the atmosphere at different scales, such as deep depressions and hurricanes. But the tornado is quite different in structure and development, as well as in size, and develops from an individual storm cell. So, it is possible to define a tornado as *a strongly rotating column of air fully extended between the ground surface and its rapidly growing parent cloud.*

By itself moving air cannot be seen, although its effects are often visible. It is almost impossible to see a tornado without these effects. The two principal ones are swirling debris from the ground and the tall column of cloud produced by the convergent, rotating air. From the surface, airborne materials vary from the natural, such as plants, soil and water, to the loose debris from shattered structures that have been unable to withstand the violent winds. These can produce spectacular effects as they may be lifted as high as the cloud base and even higher. But the larger, more visible debris may rise typically to about one third of the ground to cloud distance. In many parts of Ireland, much of this debris cloud may be obscured by the hilly terrain, unless an observer is very close. The total effect can be a dark swirling mass of materials which may cause more damage and injury than the winds themselves.

The second effect is the spectacular column of cloud produced as the falling air pressure in the column's centre causes the condensation of its water vapour. This is usually the most visible feature of the tornado. It can be seen from considerable distances since normally, but not always, it extends from the cloud base right to the ground. The condensation funnel may be much narrower than the rotating debris zone which, at the ground surface, is closer to the true diameter of the tornado. An extremely rough rule of thumb is that the visible funnel may be up to 10 per cent of the diameter of the full rotation. However, this varies because of the different amount of water vapour in the air beneath clouds. When the air has a large supply of water vapour and the pressure drop is large, the funnel will also appear to be very large. In fact, it will appear much larger than if the same tornado developed in less moist air. So, appearances can truly deceive, and it should always be assumed that a tornado is much wider than its appearance suggests (Figure 5.2). In drier environments it is possible for a full vortex to develop from the cloud base to the ground with relatively little condensation. But in Ireland the relative humidity is usually high, especially near the ground, so the vertical extent of the visible funnel will be close to that of the full vortex. The width of the funnel may be less so.

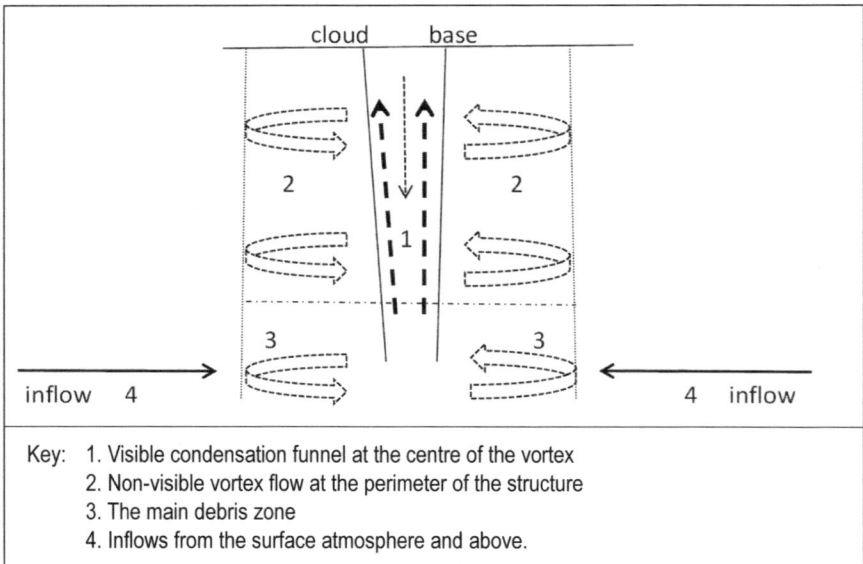

Key: 1. Visible condensation funnel at the centre of the vortex
     2. Non-visible vortex flow at the perimeter of the structure
     3. The main debris zone
     4. Inflows from the surface atmosphere and above.

5.2 The visible and invisible parts of a tornado (simplified).

The pressure difference between the outside and the inner core of the funnel is greater than for any other weather system. Typically, there may be a drop of about 100 hPa. In contrast, the lower pressure at the core of Atlantic depressions crossing Ireland would be about 15–20 hPa less than their surrounding atmosphere. While there have been no direct measurements of pressure across a tornado vortex in Ireland, the velocity of the vortex winds has been assessed, from which pressure estimates can be derived.[2]

The condensation funnels of tornadoes come in many different shapes and sizes. They are particularly variable in small to medium sized tornadoes. Such variety is a product of the very differing patterns of upward and downward air flows. Some of this variation relates to the different stages of development that the tornado is passing through at any one time. It is curious how many discussions and textbook representations assume a vertical or near-vertical column. But a large number of condensation funnels have pronounced axial tilts or have a mixture of rope-like and corkscrew-like shapes with one or more relatively inclined sections. As spectacular as these features would appear to be, they are often frustratingly hidden by a dense curtain of wind-driven swirling rain. Many tornadoes in Ireland are rain-wrapped. Consequently, visibility is desperately poor, and the tornado is an even greater threat as a result. Being obscured by rain, the warning that a clear sight of the funnel might give is not available.

Surface processes

There may be more going on at the surface than meets the eye. On 17 August 1999, a tornado came to Killeshandra, County Cavan. A group of people watched the remains of the tornado's condensation funnel 'withdrawing' (the word used in several eyewitness reports) upwards back to its parent cloud base. At the same time they noted that on the ground there was still a very strong wind driving small debris from trees in a broad circular pattern. Only minor damage was being caused. But, in fact, the rotating winds were still part of the tornado's vortex. The intense circulation had weakened, and its central pressure was recovering. As a result the water droplets that had formed the lower part of the funnel had evaporated in the boundary layer near the ground surface. The watchers saw this withdrawal and thought the tornado was over. However, there was still a tight wind circulation near the ground, which was now visible only by the swirling debris. Its potential to harm remained.

The appearance of the funnel near the ground can be highly variable. This creates difficulties in defining exactly when a tornado forms or terminates. Although the visible condensation funnel lengthens as the tornado intensifies, it does so within a pre-existing vortex which may already be in contact with the surface. At this stage the surface circulation is usually relatively weak, but in most cases its intensity will then increase. So, there is more going on than what is usually seen. It has been estimated from the experience of storm chasers in the USA that if a condensation funnel is 50 per cent or more towards the ground surface, then it is likely that the air on the ground is already rotating. This means that the wind field is already that of a tornado. However, this reasoning can never be used to confirm a tornado. Direct evidence of the circulation on the ground, either by damage effects or lofted debris patterns, is required for that. Therefore, although the terms 'descent' and 'touchdown' are a simplification of a more complex process of development, they will remain with us for a long time because they are immediate and visible.

If there is no vortex circulation in the wind flow at ground level at all, then the event is not a tornado. Above the ground surface a funnel cloud may form. This is a distinctive cloud type in its own right. In Ireland, such features have frequently been called 'mini-tornadoes', because of their shape. However, this is just one of several popular uses of that term. Its varied use has led to confusion. It is hoped that, with better information, this popular practice will give way to a more accurate and meaningful terminology. In Ireland, such funnel clouds occur when some of the potential for producing a fully developed tornado is present. Indeed, when they occur over high terrain, they can become tornadoes in their own right. Such an event was the tornado on the Galty Mountains on 6 July 2004.

That day had not been a particularly windy one in County Tipperary, but during the afternoon humid air from the south coast fuelled

5.3 A tornado funnel over the Galtymore mountain range, County Tipperary. (Photo by Simon Woodworth)

thunderclouds over the Galty Mountains and produced a thunderstorm at about 15.40 UTC. A little later when forester Dan Lynch left the Coillte office in Cahir and headed along the R668, he automatically scanned the forested slopes because lightning could always trigger a forest fire. As he did so, he was shocked to see a tornado high on the upper slopes. Its rotation was evident from the small changes in shape and the cloud fragments moving around it. At the same time, further to the west, silage cutters paused only briefly to watch from a distance of 7 km, taking advantage of the dry, bright conditions away from the mountain's stormy skyscape to complete their task. Other observers were further away. Brendan McSherry, a heritage officer with Tipperary County Council, saw it from 25 km to the south-east. Northwards, towards Cashel, the tornado was spotted and photographed by Simon Woodworth from a considerable distance (Figure 5.3).[3]

The Galty range reaches a height of 919 m at Galtymore. This is the highest of a series of smooth hump-like summits on a mountain ridge that stretch from east to west for 25 km. To the north and south are lowlands which are only 50 to 100 m high. To extend itself to the low ground from the cloud base, which was close to 1000 m, the funnel would have been many times its actual length. It is unlikely to have been able to do that. So what would otherwise have been a funnel cloud probably achieved full tornado status because of the terrain. Subsequent site investigation on the upper slopes identified fresh tree damage near the upper tree line caused by the tornado.

Single and multiple cells at the surface

In Ireland, evidence from numerous tornado events has produced what appears to be conflicting pictures of how the wind behaves as the tornado ploughs its way across the landscape. Only a few people have actually been caught in a tornado as it has passed and have been able to describe something of the wind's behaviour. In Ireland, most tornadoes appear to be single cell vortices with upward motion in the centre. These are associated with strong inflow winds that can themselves be hazardous. If these penetrated into the core of the vortex they would tend to fill the low pressure vortex core and quickly weaken the tornado. That this does not occur demonstrates how the inflow of wind becomes part of the rising circular motion around the core. In numerical models and laboratory simulations a central downdraft has been observed descending through the core of the vortex to the ground surface. In the lower boundary layer this is known as the nozzle effect. The maximum velocities occur in the minimum cross-section of the nozzle as the nozzle thrusts downwards. As a result, the maximum velocity of tornado winds is located towards the ground surface. This pattern is counter-intuitive since normal vertical wind profiles show wind velocities to increase through most of the troposphere. But the tornado funnel is different. The maximum generally occurs some 50 to 100 m above the ground.

An event with such a downdraft occurred on 10 July 1983, near Roundstone, County Galway, when a strong downdraft in the central core of a tornado vortex reached the sea surface off Bertraghboy Bay. After a boat with two sailors was battered by the wind it was pressed down underneath the water in quite an unexpected manner. As the boat was submerged, the sole survivor recalled the sea being very flat.

The development of the nozzle has two consequences. Firstly, the outer limits of the vortex are pushed further from the core and the whole structure widens. Secondly, the changes in air flow are quite dramatic. The surface winds begin to diverge outward to create one or more additional smaller vortices within the wider structure of the tornado.

These have become known as suction vortices. They progress forward with the larger parent tornado circulation, while at the same time they move around its core. As a result, the actual damage track can appear very confusing when first investigated.

Evidence for two or multiple cell events in Ireland is extremely limited. Where they have been studied in the USA, they have been associated with strong tornadoes that have lasted some considerable duration. But there have been events in which some of the features of a developing multiple cell environment have been present in Ireland. In the late afternoon of 30 September 2001, this may well have occurred in County Westmeath. Subsequent detailed site investigations of the damage tracks showed that three separate tornadoes occurred simultaneously within a few kilometres of each other north of Horseleap, as well as a fourth one some miles further north near Mullingar. It is likely that the three most close to one another were such suction vortices.[4] Their paths were typically erratic. They were mostly surrounded by torrential rain, so that few people caught sight of them. However, so severe was the rain that it halted the traffic on the Dublin to Athlone road and, of course, drove potential eyewitnesses indoors.

Inflows at the surface

Close to the vortex there will be an area dominated by a strong flow of air towards the central vortex. The wind speeds involved normally vary enormously. The most significant part of this flow will be one or more inflow jets of extreme speeds. These are dangerous places. Here objects may be dragged towards the tornado, debris may be created and relatively loose surface materials (soil, water, woody litter) carried towards the updraft, thereby concentrating it. In fact this can be where the debris cloud is most dense. The distances over which such inflows occur may not be very great because they are often sourced from downdrafts associated with the storm cell itself.

The inflow towards the funnel core of the tornado has been remarked upon in many eyewitness accounts. On 10 September 2010, in Robertstown, County Meath, Linda Hynes was startled by a roaring noise that seemed to come from every direction. On looking out of her front lounge window she was amazed to see the garden boundary hedge being flattened by wind (among other severe wind effects). At that moment she was unaware that a tornado was wrecking the garden behind her house. But the house itself was largely unaffected. She was seeing a strong inward flow some tens of metres outside the main vortex core. On 5 February 2008, a similar effect was observed by Michael Peters from his home near Kinnegad in County Westmeath at the greater distance of several hundred metres, as a fully developed tornado passed during a thunderstorm, the inflow flattening grass and scattering garden objects.

Such inflow is one of the key ingredients that enable the tornado to continue. It also affects the fluctuating intensity of the tornado. If the inflow decreases, the tornado's intensity will also decrease. Likewise with regard to an increase in the inflow jet, there will be a commensurate increase in the intensity of the vortex.

Where tornadoes come from – the parent cloud

Every tornado has a parent cloud. These can be spectacular, but often they are not clearly seen by eyewitnesses. At times the weather is so bad that the tornado vortex is largely obscured by torrential rain or even hail. On other occasions eyewitnesses have been so excited at seeing the tornado they have taken little note of the rest of the sky, other than to see whether there is another funnel around. Some of the few people in Ireland who have witnessed the development of a tornado have spoken of a massively dark underside to the parent cloud, sometimes with a base described as boiling, while some have said that before the appearance of the tornado funnel the cloud base developed a slow rotation. The impression that something awesome is about to happen is a frequent one. This is particularly strong when the cloud is approaching the observer!

Forecasting a tornado would be much easier if all tornadoes were produced by one set of conditions and one particular cloud type that could be easily identified. But that is not so. Tornado parent clouds are of several types. There is the classic cumulonimbus cloud – the massive vertical cloud, usually with an anvil-shaped top, and frequently associated with a thunderstorm. As this develops to the stage where a tornado is produced, major changes take place. Some of these cannot be seen because they occur within the cloud, but one that is visible is

5.4 Cumulus congestus with a funnel cloud in Connemara. (Photo by John Armstrong)

the emerging wall cloud. It appears late in the development of the storm when the clouds are darkening. Part of the cloud base may descend, appearing like a circular wall with an apparent slow rotation. This is the wall cloud. It was so named by Theodore Fujita who, in a pioneering study of tornadoes near Fargo, South Dakota in 1957, collected photographs, slides and films of a storm's cloud base.[5] He found that the wall cloud was formed by rain-cooled air being pulled into an updraft from below. This air is already near saturation, so when it surges upwards into the updraft, where further cooling leads to condensation, it makes the cloud above appear to grow towards the surface. However, watching one in hope and expectation of a tornado may be disappointing. Wall clouds may occur without producing a tornado at all.

In contrast, other cloud forms seen are not as spectacular. They will show vertical growth, but when the tornado appears they are usually still growing and internally organised in a less complex way than the cumulonimbus. Normally these are not thunderstorm clouds and may not even have any significant precipitation falling from them. In particular, in Ireland, rapidly developing cumulus congestus has been frequently associated with the tornado (Figure 5.4). This is a very common cloud in the Irish skyscape at other times as well, so its presence alone gives no indication that a tornado may occur. All it signifies is that there is moisture in the air and that significant upward growth is taking place. This is just one of several key processes that contribute to the creation of a tornado.

Tornadoes, Thunderstorms and Supercells

Many, but not all, who have seen a tornado in Ireland associate it in their recollection with a thunderstorm. Those who do not make that association need not blame a faulty memory. A significant number of tornadoes occur in non-thunderstorm conditions, so it would be misleading to give the impression that they are only a product of thunderstorms.

Another distinction that has been made is between the normal and the supercell tornado. The former, often referred to as a non-supercell tornado, may be either single- or multiple-celled. It has become a practice among many to classify tornadoes as either supercell or non-supercell events. But it seems odd, and far from normal scientific practice, to classify the overwhelming majority of tornadoes on the basis of what they are not. Estimates in the USA, based on the most extensive and comprehensive data set available, suggest that supercell tornadoes constitute only about 10 per cent of tornadoes at most. It is likely that Ireland experiences a similar low percentage.

Generally the largest and most damaging tornadoes are of the supercell variety, even though they are much less frequent. While this distinction reflects differences in structure, development and intensity, it is always true that tornadoes of all kinds are very damaging and dangerous. The distinctions are more important to forecast meteorologists and scientists rather than to people exposed to their threat. The Irish tornado record shows that both types occur on the island. But the majority are normal tornadoes and are produced by a variety of weather environments.

Because supercell tornadoes have caused the greatest loss of life and material damage, a large proportion of research has been focused on these. But the ordinary tornado still causes significant damage, and its winds often reach in excess of 160 km/h. Thus, they are a severe threat to human life. But, having been researched less, they are not understood as well. It seems clear that some of the conditions necessary for each type of tornado vary in a number of respects.

The significance of boundaries

The birth of a tornado is mostly about airflow boundaries in the atmosphere.[6] Irish tornadoes have been found to be initiated largely by vortices produced by changes in direction and speed of the wind at various altitudes through the lower part of the atmosphere, known as the troposphere. Such wind boundaries have the potential for developing a tornado if there is convergence in the airflow, spin and accelerated vertical growth.

Weather boundaries that separate successive weather systems and different features on a synoptic map presented on the TV or other news media are familiar to most people. We are less familiar with the important boundaries that occur with respect to wind direction and speed away from the ground surface. They are often visible when the clouds above us are seen moving in a different direction to the wind around us on the surface. For example, it is not uncommon in Ireland that ground-level southerly winds are overlain by more westerly winds. Wind shear occurs where the transition between the two flows occurs. In this case it is directional shear. It is usually the case that the higher altitude winds are also faster, a condition that produces speed shear. Such changes in wind speed and direction may occur at different levels in a vertical wind profile. If the shear is large and the air is forced to move from the lower to the upper level it may acquire a spinning motion (vorticity). However, initially these non-convective vortices may not be very deep.

The next required ingredient is a stretching mechanism.[7] This is created by what is going on both below and/or above the zone of marked wind shear. Below it, converging airstreams may increase the uplift and intensify the vorticity. Sometimes, zones within weather systems may develop this convergence as airstreams are either drawn or pushed together by differences in air pressure. But surface conditions are also important, especially in Ireland. Examples are when increased friction creates convergence by slowing down air as it passes from the sea to land, or when the topography channels different surface airstreams towards one another. Low-level convergence within airstreams is very common in Ireland and contributes significantly to storm cell development when other conditions are also favourable.

Convective instability, when air is heated from below and rises, may have a similar effect. One measure of this is made whenever upper air soundings are taken, expressed in terms of the convective available potential energy (CAPE).[8] Research in the USA has shown that where such convection plays a significant role in the development of a tornado, CAPE values would normally be above 2500 J/kg. But typically in Ireland instability indices are generally very low, even in storm situations. The instability that does occur tends to be located in the lower part of the troposphere. Thus its role in most cases appears to be to give a small extra boost to the convergent airstreams as they surge upwards.

Air that is rising because it is warmer than the surrounding air will only continue to do so if it remains relatively warmer. It is not unusual for this zone of instability to come to rather an abrupt end when it is overlain by another air stream that is slightly warmer at that altitude. This may occur because its own journey to Ireland has given it different characteristics. In this way, the low-level instability may be capped by a temperature inversion, like a lid on a saucepan. When this cap is eroded by the build-up of energy from below, explosive upward growth will result. This can provide the required strong stretching mechanism.

A further process that aids strong vertical cloud growth is moist convection, occurring when lifted air cools and its water vapour condenses. The resulting water droplets show in the massive build-up of cumuliform clouds. As this condensation occurs energy is released, thereby further fuelling upward growth of the storm cell. The term 'moist convection' is used to describe this process because the moisture provides so much of the necessary energy to make it happen.

Upward stretching may also be related to the very high wind speeds that can occur at a high altitude from time to time, especially when the winds at the upper level are divergent. A strong flow in the upper troposphere (e.g. at the 300 hPa level, which is the level often used to monitor the upper jet stream) may produce intense upward movement and storm outflow at this higher level, despite the modest thermodynamic profiles which show only weak potential instability (e.g. low CAPE values) beneath them. This is similar to a freshening wind increasing the draw of a living room fire as it passes across a chimney. A similar consequence may be produced when the upper level wind flow diverges (i.e. spreads out) at the same level. This may stretch the column of air beneath it as it seeks to replace the divergent air above.

The Banemore tornado

The way these ingredients typically come together to produce a tornado can be illustrated in the case of the Banemore tornado of 3 July 2000. Banemore is in County Kerry, near Listowel. It is only 70 km from Valentia upper air station, so there is reasonably good data available to describe the atmospheric conditions that prevailed close to the time. In addition, good photographs and eyewitness accounts are available since it occurred in the later afternoon in good visibility (Figure 5.5).

There were several key ingredients to this event: wind shear between different altitudinal levels, local convergence of the surface wind streams, and a small amount of uplift induced by the rising terrain. Above these

5.5 Photo of the Banemore tornado, County Kerry. (Photo by Tim Griffin)

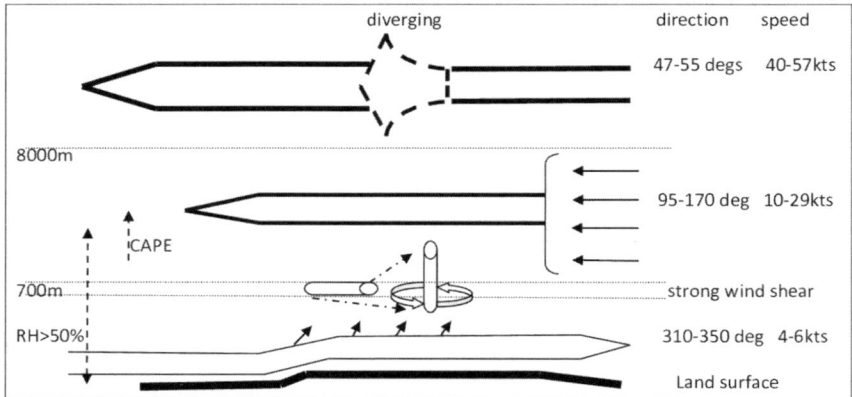

5.6 Ingredients that produced the Banemore tornado.

lower-level conditions there was a zone of divergence in the upper level wind. In addition there was a deep layer of moist air to fuel some convection and a very minor amount of convective instability (Figure 5.6).

The most striking feature of the event was the strong wind shear conditions that prevailed at different levels of the atmosphere. Between the surface and an altitude a little above the upper slopes of the Kerry mountains the wind changed direction from north-north-west to east-south-east. This amount of directional shear is strong for such a thin, shallow layer. In contrast, the wind speed shear was relatively slight, from 5 kt to 10–15 kt. But at higher altitudes this changed. Through a much deeper layer, to 9170 m, wind speed shear was more marked, reaching 50 knots, while the continuing directional shear went through another 40 degrees to the north-east. So overall the potential vorticity was strong.

At the higher levels, the strong north-easterly wind flow was not consistent in direction. Over south-west Ireland it was beginning to diverge in a fan-like pattern between south-west and south, spreading more broadly further away from the region. Above County Kerry, this divergence was already making itself felt and became a contributory factor in the upward stretching process above the sheared condition of the wind profile beneath.

At the surface there was some convergence in the airstreams that forced uplift from below. Throughout County Kerry the surface winds were light. The centre of a tight low pressure system lay off the south-west coast. The tight curvature of its core gave south to south-easterly winds in the south of the county, which were mostly either partly blocked by the south-western mountains or lifted through and above them. They then converged with the winds arriving from the north-west and lifted, adding to the uplift already being forced upon the low pressure system by the higher terrain. In addition, it is possible that the outflow wind from an earlier thunderstorm may well have strengthened convergence locally.

As the forced uplift cooled another factor came into play. The upper air sounding taken at Valentia shows that the air was relatively humid, to the level where the atmospheric pressure was 800 hPa at 1990 m. At this height the winds had already become easterly. The cooling experienced by this moist air as it was lifted over the high ground not only produced water droplets to grow the clouds, but thereby it released energy into the growing storm cell helping to fuel further upward growth.

There was a little convective instability to add to this. This was confined to low levels and was extremely small. On this occasion over County Kerry the CAPE value for this layer was approximately 3 J/kg, which is relatively insignificant. While locally this may well have been greater, it would still have remained of minor significance.

Thus, a combination of low-level and high-altitude factors gave rise to an environment that stretched the vorticity created by the wind shear through the lower and middle levels of the troposphere. At the lowest levels, convergent air flow combined with the uplift provided by the surrounding mountainous terrain became the predominant means of creating strong vertical growth in the air column, with the added benefit of whatever minor instability was available. This coincided geographically with the high-altitude divergence within the upper north-easterly air, to produce a pull-up factor for the stretching process. So, the initial vortex produced by the shearing environment was able to spin further upwards, then stretch and intensify in so doing.

Supercell tornadoes

In Ireland, the supercell tornado may be considered as a special case. Such tornadoes are much less frequent and have a particular structural feature compared with other tornadoes. This is the single deep, rotating updraft that forms the mesocyclone at the core of the storm cell. It surprises many people to learn that the supercell was first identified and defined in the UK rather than in the USA.[9] The original reason for using this term was that the largest, longer-lasting thunderstorms that produced the most severe tornadoes were unicellular, consisting of a single organised storm cell.

By the time of this development in the UK, research in the USA had already identified the key conditions for the formation of tornadoes as high instability (CAPE), strong moist convection and marked wind shear.[10] For a supercell, a large amount of moisture is involved in producing a massive unstable parent cloud, even without the intense precipitation that also normally occurs. Indeed, the more moisture there is in the lower air, the larger the condensation funnel that may eventually develop at a later stage. The moist convection is strengthened, and its life is lengthened when the updrafts and downdrafts are separated. This prevents cold downdrafts falling onto, and thereby suppressing, the updrafts. The successful separation of rising and descending air is achieved through strong wind shear conditions as these ensure that falling rain occurs away from the surface inflow.

The interaction between inflows and outflows has another important consequence. Along the axis between descending, outflowing air on the one hand and ascending, inflowing air on the other, a spinning action is generated. This must then be tilted towards the vertical by the strength and positioning of the updrafts and downdrafts. As this spin (vorticity) is repositioned to a more vertical alignment, its lower end may draw in more converging air from lower levels, further stretching the vertical dimension of the rotation. Thus is created a large rotating updraft within the cloud, known as the mesocyclone. Usually it has a large diameter of many kilometres.

It should be stressed that the supercell is not determined by the depth of convection in the cumulonimbus cloud, even though the convection is normally very deep. Some accounts of the supercell define it as having one updraft and downdraft instead of several that interfere with each other. But deep convection can occur without the rotation that is an essential feature of the supercell's structure. It is misleading to think that the supercell is simply an exceptionally large convective cell in a well-developed cumulonimbus cloud.

The rotating updraft section of the supercell was identified in 1959 by Fujita in the USA and it was he who first used the term 'mesocyclone' with which it was associated.[11] A few years earlier the hook echo and rear flank downdraft had been identified.[12] The hook echo became regarded by many as the signature of the supercell tornado. The rear flank downdraft was an associated feature. It was found to create a small cloud-free slot as the downdraft descends before wrapping itself around the area of the storm where the tornado would tend to develop a short time later.

Within the supercell's rotating mesocyclone a narrower vortex is created by an area of increased spin, as part of the updraft stretches and intensifies at the same time. In the section of the vortex beneath the cloud base the central pressure values fall and the vortex intensifies there as well, although it may not be visible initially from the ground. This process produces more condensation of water vapour creating more cloud droplets. These make the vortex visible as it expands towards the ground. It appears to grow downwards because the atmospheric pressure beneath the cloud is already lower than at the ground and so it does not have to fall as much to achieve condensation, compared with portions of the vortex closer to the ground.

Downdrafts from middle levels of the troposphere have also been found to make a significant contribution to the development of some tornadoes, particularly supercell tornadoes. These originate from middle-level flows outside the storm itself. They become part of the storm as they collide with it. As a result, some of this flow deflects towards the surface, bringing drier air downwards into the rear portion of the storm cell. In so doing, its dryness enhances evaporation along its track and in some cases clears a path through the ice and water particles that make up the cloud. Such clear slots are frequently observed by storm chasers to position themselves in anticipation of tornado development or intensification.

The evaporation further cools an airflow that is already relatively cold from its relatively high altitude origin. Such descending air rapidly becomes much denser, accelerating the flow, typically reaching speeds of 60–80 knots. Any horizontal vorticity present will then be tilted towards a more vertical alignment and is likely to be stretched towards the ground. The downdraft also contributes to stretching vortices upwards, since they not only spread out at the surface, but surge upwards as well, as they rebound from the surface. These flows enter the tornado from the rear and, as they converge with the vortex, they accelerate its rotation. This process also helps to create the hook echo on radar images recorded during such events.

A variety of supercell types have become recognised, although the fine distinctions between them are not very readily noticeable in a small area and limited environment such as Ireland. The basic one has been termed the classic type, which has also had a miniature version attached. In addition the high precipitation (HP) and low precipitation (LP) types are categories frequently used, not least by the chasing community. But in reality, rather than discrete types there is a continuum of storm identities that merge into one another.

Classic supercells typically exhibit a textbook appearance. From the ground, low-level storm features that can be seen include a wall cloud, a rain-free base, a rear flank downdraft and distinctly separate updraft and downdraft regions. They can be relatively isolated. Their radar features may include the presence of a hook echo and other characteristic weak echo regions. The miniature version of these has been associated with limited buoyancy to only a moderate depth in the atmosphere. An area of very dark skies, then intense precipitation (sometimes with hail), tends to occur in the forward sector of the cell, ahead of the rain-free base. As the mesocyclone rotates anticlockwise it may draw the rain and hail out of this forward area towards its left flank behind the wall cloud, creating a curtain that can obscure the wall cloud from that side, but which is often not very evident from the right flank.

HP supercells are usually less isolated than their classic counterparts and often form in environments with a much higher degree of atmospheric moisture and significantly weaker mid-level storm-relative winds. As a result, the separation of rain and updraft is less successful, so the updraft tends to become rain-wrapped. This makes it difficult to identify low-level storm features such as wall clouds. In particular, HP supercells can hide tornadoes, making them very dangerous. The precipitation is not just greater in total, but is much more widespread beneath the supercell. Heavy rain and hail may occur from the forward left flank around the left side to the rear of the mesocyclone.

LP supercells occur where there is much less atmospheric moisture but strong mid-level storm-relative winds. This means they do not produce much precipitation and cloud features such as wall clouds can be much less marked. However, when there is precipitation, it can be very hazardous since it often takes the form of very large hail. All this is very visible not just because of less precipitation, but because the cloud base is often somewhat higher.

It is difficult to generalise about the height to which the vortex ascends within the storm cell. In some circumstances this can extend up to and beyond the upper surface of the storm cloud. This has been observed frequently by storm chasers viewing storm cells from a distance, who have described the bulge that sometimes develops in the upper boundary even when there are strong horizontal winds above the cloud tops. Many rapidly growing storms have this feature, but it tends to have a very limited duration, as it grows and then declines. However, when viewed from a distance, this upper dome lasts for more than ten minutes or so, so it is usually reckoned to be part of a developing supercell.

A supercell tornado at Ballysadare

Such a supercell occurred in County Sligo on 6 August 1999.[13] Reference has already been made to the role this probably played in raising Irish public awareness to tornadoes. It was first seen by a number of Ballysadare residents as they watched storm clouds moving towards them from over the steep slopes of the Slieveward Hills to their south, about a kilometre away. A period of intense heavy rain had stopped and in almost calm conditions their attention was drawn to an extremely dark cloud appearing across the skyline. The storm cell showed the presence of a mesocyclone, although this was not something the anxious residents could appreciate. In the darkening light they saw the mesocyclone's lower parts slowly rotating in a broad sweep almost above them, while its relatively wide underside was described as boiling and raging in a most unfamiliar way. From this a funnel descended quite quickly, 'within a minute or so', it was reported. It then spun away along a track towards Streamstown and Ballysadare Bay in a cloud of spray from the saturated ground (Figure 5.7).

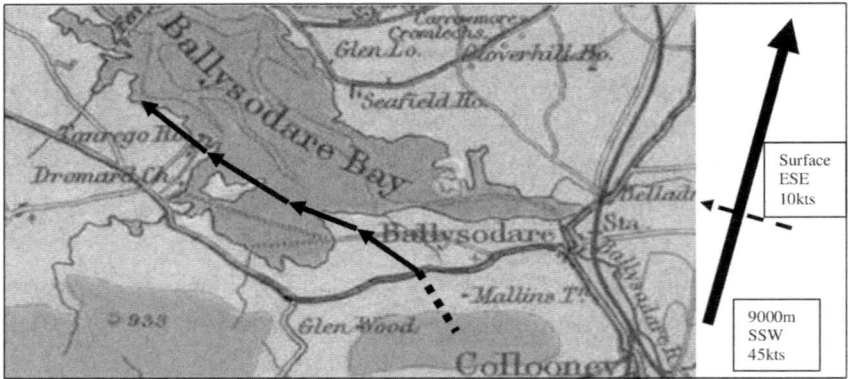

5.7 The Ballysadare tornado track showing the wind shear directions between the upper air and the ground surface, the funnel's descent (dashed line) and the track of the tornado (arrowed line).

On the opposite side of the bay, the tornado was caught on video. It captured the base of the mesocyclone as well as the details of the narrow condensation funnel. Although this appeared to be only about 10 m wide at the surface, a much wider surface rotation was evident from the spray droplets swirling around it. The wider base of the supercell in the sky above was small, only 2–3 km in diameter. However, large size is not a necessary condition for a supercell, despite the impression given by the term. Indeed, it is likely that most supercell tornadoes that occur in Ireland are small.

Subsequent examination of available meteorological data showed that this was indeed a low-topped single cell storm. The amount of directional wind shear was outstanding. By the afternoon there had been a dramatic shaping of the vertical wind profile which had near easterly surface winds but at 9000 m they were from the south-south-west (Figure 5.7). There was also a doubling of wind speed between these levels, to 45 knots, which is not uncommon in storms. The directional change of more than 90 degrees, providing significant potential for spin, occurred mainly in two steps. The lower, main step was at 2400 m and coincided with a temperature inversion. This feature limited vertical air movement above it until it could be eroded during the afternoon. This happened as it passed over the warming surfaces of inland Ireland. The sudden explosiveness of this when the barrier was finally broken not only allowed marked warming to spread to a high level and thereby increase the instability of the vertical air profile, but the dramatic release also helped to stretch the upward current, gaining spin as it did so from the changing wind direction above. In addition, the doubling of wind speed encountered at this higher level played a significant part in further stretching the upward trajectory, helping to pull the trigger that caused the tornado. So while the local air temperatures in Ballysadare were nothing out of the ordinary for a summer afternoon, the transformations taking place as the air journeyed towards Ballysadare town were to create an exceptional experience for the residents there.

Expanding the Definition of a Tornado

A slightly fuller definition of a tornado can now be developed that incorporates the wider range of distinctive processes and features that have been outlined. It is as follows: *The tornado is a vortex of strong winds that stretches between the ground and cloud level, associated with an updraft into a cumulonimbus, towering cumulus or line squall and is usually accompanied by a condensation funnel within that vortex that may or may not reach the ground. The vortex winds are usually violent and damaging and generally the vortex produces a distinctive roaring-like noise.*

The different meteorological conditions associated with tornado activity are such that they include a range of conditions that can reasonably be expected to occur over Ireland a number of times in any one year. One major theme of research over many years has been focused on the highly varying conditions within and along a cold front. Some recent work has used Irish and UK data and may well make an important contribution to forecasting tornadoes here in years to come.[14] But this is just one of several environments creating our tornadoes. This means that we have no reason to assume that tornadoes are, or have ever been, strangers to the country. Indeed, we can expect both the classic and the supercell tornado to occur each year as a normal part of our climate, even though we cannot be precise about their relative importance. Many of the more recent advances in detecting and understanding the differences between tornadoes and the environments that produce them have come from applying developments in radar technology and in computer modelling.[15] These techniques are now beginning to be applied to Irish tornadoes. However, observation is still an important tool for advancing our understanding of these events. In this area considerable work is being done in Ireland and forms much of the basis for our understanding of variations in the types of tornadoes that occur each year.

As more detailed studies have been carried out, using new and more developed field methodologies, so our understanding of the tornado has become more complex. The increasing activity and detailed recording by storm intercept and tracking teams, together with more sophisticated cloud and weather modelling, have led to the recognition of different types of tornadoes (e.g. landspouts) and other dangerous vortices (e.g. eddy whirlwinds and gustnadoes). As our understanding of these has grown, so we have been able to determine and define the particular tornado profile and tornado-producing environments in different regions of the globe. It cannot be assumed that the research results generated in one part of the world are sufficient to explain what happens in others, even though there will be many close similarities.

CHAPTER SIX

Waterspouts on Irish Waters

For it haps that sometimes from the sky descends
Upon the seas a column, as if pushed
Round which the surges seethe, tremendously aroused.[1]

Waterspouts have long been recognised and for centuries they terrified the most experienced sailors. They were familiar to the Romans, for the poet Lucretius described them in his epic poem *De Rerum Natura* (above). It was not just the seas around Europe where a literary record built up over the centuries, but also some of its inland water lakes where they became part of local folklore.

Ireland also has a long history of waterspouts. The earliest traceable record probably goes back to the account of the Irish saint, St Abban.[2] It tells of an event when Abban saw a monster close by on the sea, observed from his ship. The monster was described as having a hundred heads of diverse forms and as many ears. Significantly, it was also described as extending itself into the clouds and creating such a turmoil in the sea that the ship was in danger. Many features of this story point to the high probability of this being a waterspout. St Abban himself was the son of King Cormac of Leinster and founded many churches in County Wexford, probably including one on Our Lady's Island on the Wexford coast, and served as the abbot of Killabban Abbey until 16 March 620. This places the 'probable' waterspout event in the early seventh century.

A small number of other records of water monsters are found in a variety of early texts, but few give sufficient evidence for them to be seriously considered as possible waterspouts. Nevertheless, Bishop

William Reeves (1815–92), a former president of the RIA, recorded how certain rivers and lakes were thought to be haunted by serpents at a very early period and that this was still a contemporary belief in the nineteenth century, as far as he was aware.[3] Indeed, this belief has persisted into some modern accounts in recent decades, which have reported the sighting of similar phenomena. An example is the 'wurrum' that is seen in several mountain lakes.[4]

Waterspouts were sufficiently significant to become part of the Irish language. The general Irish words for whirlwind and land-based vortices, such as the *sí gaoithe*, were also used when a waterspout was reported and a record made for posterity. But the waterspout was also called *cuaifeach uisce* and *maidhm bháistí*.[5]

By the nineteenth century the waterspout had been largely demythologised. So, for example, the report of a large waterspout on Lough Neagh on 25 August 1872 was made in terms we would recognise today. This appeared during a thunderstorm over the lough and moved rapidly northwards over the lake surface. It came ashore south of Staffordstown, County Antrim. It then progressed inland as a tornado where it created a track of damaged trees and rooftops that extended for at least 5 km.[6]

Changing Terminology

However, the use of the term 'waterspout' has changed in Ireland. Historically, the earliest use was for a phenomenon that was exactly what the word indicates – a spout of water pouring from the clouds. When it hit the water, naturally a large volume of spray would be thrown up. This idea lasted a long time. It was challenged during the eighteenth century and became the subject of a debate among scientists. This traditional 'stream-of-water' theory was energetically defended, and it took time for the concept to change. It was deeply rooted in Ireland, as demonstrated by one of the great local weather tragedies of recent centuries. It happened at Glenflesk, County Kerry.

The Glenflesk waterspout

Humphrey O'Sullivan's diary (to which reference was made in Chapter 2) has a record of the event.[7] Although he was a teacher in Callan, County

Kilkenny, he affectionately recalled in his diary that he had been born and bred in Gleann Fleisce (Glenflesk) near Killarney. He recorded a waterspout having occurred on 16 August 1831. Unlike Killarney, Glenflesk is not a lakeside settlement. It is a small village in the upper Flesk valley, which drains the high borderlands between counties Cork and Kerry. During an intense thunderstorm, a highly localised torrential downpour in the mountainous terrain was reported to have caused widespread damage and the destruction of local farms. Some more modern interpretations have suggested that this was a tornado, mistakenly called a waterspout. But wind was not a feature of contemporary reports; these were all about the torrential rain. Many people died, including eight from one house swept into the nearby River Flesk, in addition to losses of cattle, corn, turf and all the crops.

Such waterspout events are recorded from time to time in the Irish historical record, thereby muddying the water when trying to construct the historical profile of tornadic and waterspout vortices. In the same manner, historical waterspouts were recorded by Armagh Observatory at English Town and Lower Eglish on 29 July 1805, while on 27 September 1811 it was noted that there were 'Waterspouts in the east about 4 pm'.[8] Thus, it was an established term with widespread use right across Ireland over a considerable historic period.

The Roaringwater Bay waterspout

Another tragedy, this time in County Cork, gives a more accurate reflection of the modern meaning of the term 'waterspout' and is found in much more recent historical literature. On 5 December 1908, the lobster boat *Water Lily* was in the Long Island Channel of Roaringwater Bay. It was heading westwards towards the island, the home of skipper John O'Driscoll, who was with his 18-year-old son Jack.[9] Accounts of the events that followed have been passed down within the community. These recall a moderate breeze and no awareness of any apparent danger from a sudden squall. John O'Driscoll saw a second boat, belonging to another local fisherman, John O'Donovan, ahead of the *Water Lily*, so he and his son tightened their gig's mainsheet to catch it. Thus preoccupied, they failed to see a vortex rapidly approaching from behind. Jack remembered his father's shout to let go the mainsheet before it overwhelmed them. John O'Driscoll drowned, but his son was rescued by a boat put out from the island. There the waterspout and its moving mist of foam had been seen and watched with horror by some of the islanders.

The same waterspout had already caused damage in the coastal townland of Colla. There, as a tornado, it had tracked through the townland, crossed the shore and accelerated as the frictional drag of the surface decreased significantly on the smoother surface. The tornado became a waterspout and swept towards the boat, with fatal consequences for John O'Driscoll. For a vortex to make such a transition between land and water surfaces is not uncommon in the Irish environment.

Water Surfaces and Waterspouts in Ireland

As a relatively small island, Ireland imposes the sea upon its inhabitants, both as a resource base and as a lived space that exposes communities clustering along the coastline to its severest weather conditions. These include waterspouts. Beyond its coastline of 2,797 km, to its territorial limits, is a territorial sea in excess of 60,000 km^2 of water.[10] This expands the Irish surface environment of 84,000 km^2 by at least 70 per cent. Most of these coastal waters are visible from the land on a clear day and at times this is even possible when there are storms at sea.

Exposure to these coastal waters has not been a simple matter of choice for mainland dwellers as might be supposed, despite the economic, social and political imperatives that historically encouraged coastal settlement. In addition, the several hundred offshore islands were (and in some cases still are) homes of other communities, both large and small, who also navigated these waters as they coexisted with their mainland cousins. In 1841 more than 220 of the islands had inhabitants. Even though the next 150 years saw widespread decline in their populations, there were still about thirty-five occupied islands and a huge increase in the size of mainland coastal settlements and expansion of water-based activities by the end of the twentieth century. So, although the population pattern has changed, the vulnerability to water-based hazards such as waterspouts remains.

The mainland has a significant number of water bodies as well. The fifty most significant inland loughs range from the largest freshwater lake in Britain and Ireland, Lough Neagh (383 km^2), to the small Lough Bane on the border of counties Meath and Westmeath (with an area of 1.5 km^2). Unlike Lough Neagh, a good number of the loughs are much longer than they are wide, both in mountain and lowland areas. This can create a longer water surface track for a waterspout if its movement is in a suitable direction. Again, these lakes have often been significant in the settlement history of Ireland and the communities clustering around the lakes have always been subject to their severe weather hazards.

The water surfaces of Ireland have a number of characteristics that contribute distinctively to waterspout occurrences, although these will neither produce waterspouts by themselves, nor in combination. Nevertheless they will enhance the likelihood of a waterspout when favourable atmospheric conditions occur. Interactive processes between atmospheric vortices and their underlying surfaces are probably the least understood part of waterspout and tornado systems. But studies indicate that there are certain features of the surface that are relevant, particularly through their distinctive contribution to critical energy transforming processes at the surface.

Surface friction

Compared with land surfaces, water surfaces have relatively low friction. On land, urban structures and irregular terrain may reduce wind speeds by 40 to 50 per cent, but over water this is only 20 to 30 per cent.[11] The height to which this frictional drag has an effect is also much shallower over water, usually by several hundred metres. To maintain a vortex motion a continuous energy supply is needed that is sufficient to overcome the effects of friction. Where there are high levels of friction much greater energy is required. Thus the persistence of vortices on a water surface may be possible where they would dissipate quickly on a land surface. As a result of this, waterspouts often terminate suddenly when they cross a shoreline, both at the coast as well as on inland waters. The converse may also apply where a vortex may speed up and even intensify when it crosses from land to water.[12]

In addition to the generally lower frictional drag over water, it has been shown that there is much less variability in that friction as the surface air streams across the water surface. Since large frictional variability over relatively short distances is potentially disruptive to the organisation of atmospheric vortices at the surface, its lack helps in the development of a circular flow pattern that would otherwise become disrupted. Once formed, the lack of frictional variability also helps to sustain the vortex as it moves across the water surface. So, potentially, waterspouts have a longer lifespan and track length than their tornado counterparts on land.

Surface water temperature

A particular characteristic of the seas around Ireland is the relative warmth of their surface water temperature compared with the same latitude around the globe. But despite popular comment to the contrary, their convective influence is quite limited. Seasonal change to surface sea temperatures is slow because the rising solar energy input of summer warmth is spread deeply into the sea by the vertical movement of the water. Winter cooling is similarly slowed because of the deep storage of warmth that has to be dissipated to make it happen. In contrast, because land surface materials are static, thermal gains and losses are concentrated to a much shallower depth. The rise and fall of land surface temperatures is much more pronounced as a result.

Thus, sea surface temperatures (SSTs) rise and fall only slowly compared with those on land. For these to contribute to waterspout development a sufficient thermal contrast between the surface and the air is necessary. Energy transfer from the water source to the surface air will then help to achieve the level of instability conducive to waterspout development. But exceptional or intense instability is not a prerequisite for this. Indeed, they have been shown to occur under quite weak instability conditions.[13]

Warm air arriving from southern directions towards Ireland is invariably being cooled from below. This is significant since the predominant airflow across Ireland is from the south/south-west/west directions. As a result, convective processes within this airflow are dampened down. In contrast, land surface temperatures heat quickly and such air crossing the coastline may suddenly be heated from below and develop instability characteristics. The degree to which this happens depends on the highly variable land surface characteristics.

But the Irish climate is also shaped by a significant frequency of air from more northerly directions. In contrast to the southerly and westerly flows, all of these airflows will be warmed from below, both over the surrounding sea and over the land. Thereby their internal convective processes are strengthened, whatever the season. These may be dampened down on reaching land when the land surface is particularly cold, as in winter, but further intensified when land surface temperatures have significantly risen in summer.

There are some areas of offshore waters that are significantly different and can develop their own thermal character. They can impose their own modification on the warming and cooling patterns due to their relative shallowness. These are coastal waters less than 10 to 20 m deep. As a result, their SSTs rise much more rapidly than in deeper water. Recent monitoring surveys are beginning to show some such limited areas of SSTs that could warm the lower atmosphere significantly. This has the potential for compensating for the rather limited levels of convection in Ireland's predominant airflow, caused by the underlying widespread effects of the generally cooler SSTs.

However, such shallow water areas around Ireland are generally quite limited. So overall, the spread of a strong warming effect from this coastal source is slight, unless the air happens to flow roughly parallel to the coast. Exceptionally, onshore air might be warmed in this way if the wind is slack enough for the position of the air mass to remain relatively constant over the enhanced SSTs. It is also noteworthy that because Ireland has a highly irregular coastline with numerous small and large islands over short distances, localised differential sea surface heating patterns with the potential for local 'hot spots' under favourable synoptic circumstances do occur.

Localised instability may also be generated by the introduction of locally sourced colder air above sea surfaces. This comes about occasionally, when the atmosphere is relatively stable and land breeze systems develop along the Irish coast where there are mountain areas nearby. Then, the night-time katabatic winds flowing downslope across the coastline produce a layer of cold air above the coastal waters. The resulting strong temperature contrast may generate instability. But this effect is not frequent because many mountain areas either drain into narrow rias and bays, or their highest altitude is limited in area, thus restricting the amount of cooling. Also, the longevity of weather systems that would be required to make this factor really effective is not very common.

Surface-level water vapour

The shallow layer of air immediately above a water surface has an abundance of water vapour that is quite different from its counterpart over land. This may extend upwards for hundreds of metres. When the air is relatively stable vertical air movement near the surface is minimal. But above the surface moist layer, the cooler air in contact with it may cause some cooling and condensation at the top of the moist layer. The latent heat of condensation thus released is considerable over warm water. The significance of this is that energy is released and further vertical growth is possible, drawing the water vapour upwards. This is reinforced by the humid air being less dense than relatively dry air, so any vertical water vapour gradient will help to strengthen the uplift.

In this way uplift and its associated condensation of water vapour is a key process in waterspout development. It is needed for the localised vertical air currents that produce the cloud with which the waterspout is associated. The same process also supplies most of the energy to power the subsequent waterspout. When this gets cut off due to the air feeding the waterspout circulation becoming cooler the waterspout will dissipate. This is often due to nearby rain showers developing from the associated cloud system drawing down cold air to the surface.

Where Waterspouts Form: Waterspout environments

A number of authors have attempted to identify different types of waterspouts. Some of these types are based on physical properties of the spouts while others are based on the mechanisms that produce them. But most of them appear to have a number of features in common. Firstly, converging winds associated with horizontal shear lead to horizontal spin (vorticity). Then, vertical growth of that spin is linked to an associated cloud system that develops throughout the event. In these basic terms the waterspout formation processes involved in the Irish environment are similar to anywhere else in the world.

Potential trouble spots: convergence and vorticity

In analysing a weather map for potential waterspout areas the search is on for areas of convergence where winds from different directions will tend to come together and pile up. At the synoptic scale these are difficult to pick out. Every depression has some vorticity because the air spins round in an anticlockwise fashion. But at any local point this is too small to produce a violent vortex. However, there are circumstances where some of this vorticity becomes concentrated and localised spin effects can develop. This may occur when strong directional and wind speed changes occur, and wind gradients suddenly develop. A number of contexts give rise to this potential:

1. **Along a front when the wind veers and converges, or within a squall line or other vertically developed cloud systems where outflows from different cells meet and converge.**[14] These circumstances occur frequently in the Irish environment and are a principal source of vorticity in most coastal and inland lake areas. This may occur not only at the surface, but also at various levels through the atmosphere. Indeed, it is not unusual that the strongest effect is at these higher altitudes, well above the surface.

2. **Along shorelines when the wind is blowing parallel to the shore.** The frictional drag along the coast will slow the wind compared with the wind offshore.[15] For example, these conditions are frequently met along the Cork–Wexford coastline in south-west winds. They can also occur around elongated coastal islands orientated parallel to the wind. Frictional drag on both sides will tend to create convergence and downwind swirls, as well as directional deflections due to the details of their terrain. Some western and northern coastal areas have been associated with waterspouts in such environments, although it is uncertain how significant these influences are overall. It is likely that minimum island sizes and altitudinal ranges combine with critical wind conditions.

3. **From the water surface to the upper limits of the atmosphere's troposphere, when strong changes in wind speed and direction occur, sometimes over relatively short distances.** Such wind shear is often pronounced at the edge of the upper level jet stream, which is frequently in the vicinity of Ireland. But it is not confined to that circumstance alone. Horizontal pressure and wind patterns may vary at many levels through the atmosphere, and it can never be assumed that flows at upper levels are a mirror image of flows near the surface. Directional and speed variations in the airflows at different heights arriving over Ireland from different sources produce localised and regional patterns of wind shear that vary considerably in their lifespans.

4. **Along land breeze boundaries when the land breeze converges with the regional synoptic flow, especially when the regional flow is relatively weak.**[16] The land breeze is the converse of the sea breeze. By itself, this mechanism is of minor importance in Ireland because the frequency of land–sea breeze systems is limited, mostly to the July–September period. However, due to the mountainous rim close to much of Ireland's coastline, katabatic density that flows down short steep mountain slopes probably produces numerous land breeze events. But it is rare for these to be strong, and they are most likely when the mountain breeze is severely funnelled.

5. **Above shallow coastal water hot spots.** In shallow, relatively calm water the resulting heated overlying air may create a very localised surface pressure gradient and light convergent winds. However, in Ireland inshore turbulence, coastal currents and river outflows tend to work against this potential in the very limited areas where it exists.

6. **In the wake of islands and long peninsulas.** Such wake vortices are spawned on the leeward side of islands in a similar way that a rock in a stream can be seen to generate wake vortices.

7. **Above lakes in steep-sided valleys.** A strong wind gradient may develop between the valley sides and above the open water surface due to friction. This is very similar to the coastal effect.

These individual sets of circumstances by themselves would not produce strong vorticity. But it is striking what variety of circumstances may contribute to it. It is also apparent from this brief listing that convergence and vorticity may occur either at the surface or at various other levels in the atmosphere. The conditions described in 2, 4, 5, 6 and 7 above relate largely to the surface or lower atmosphere. But cases 1 and 3 are particularly important for producing violent vortices over water in Ireland. Such vorticity is a horizontal motion. For waterspouts to develop, this vorticity needs to be stretched vertically, thereby producing a vertical spinning vortex.

Realising the potential: vertical stretching

Inevitably there will be a vertical dimension to the vortex as it forms. It cannot be a simple two-dimensional process in the real environment. Convergence and vorticity alone will be insufficient to produce the fully developed waterspout. Single-level processes are insufficient of themselves to produce deep stretching. Deep stretching through the sub-cloud layer to the cloud above is required. Vortices without any attachment to clouds are rather like spin-ups that occur on street corners lifting dust and leaves and are very temporary.

Current understandings tend to distinguish between two kinds of waterspout: fair weather waterspouts and storm cell waterspouts.[17] In the case of the fair weather waterspout, stretching occurs between the surface vortex and the cloud base. In contrast, the storm cell waterspout stretches downwards towards the surface.

Thus, the necessary vertical stretching may occur in a variety of ways. There are numerous case examples in the international literature that show how updrafts within developing cumulus congestus or cumulonimbus clouds play a significant role in determining which surface vortices become stretched and which do not.[18] Those vortices at the surface that coincide with a strongly growing cumulus cell will become stretched as their positions coincide vertically above each other. For this to be successful the timing of the most strongly growing phase of the cloud must occur when it is co-located with the surface vorticity. This narrows the window of opportunity for waterspout development hugely and shows how critical are the developing conditions within the parent cloud.

However, either deep convergence from lower levels or convergence at middle levels of the atmosphere may result in vorticity at some height above the surface atmosphere, as occurs in cases 1 and 3 described above. These types may be fed and stretched by storm inflows from below and/or strong outflows at a higher level as well as by the strength of the wind shear within the storm cell. Vortex stretching then occurs within the storm cloud and will be observed only as it emerges from its base, as would a tornado if the storm was developing over land. Thus, yet again, processes within the parent cloud to which the spout is related are critical to its successful development. This is the most difficult part of the storm to study because it is often buried in a thunderstorm or shrouded by rain. The storm cloud itself will not be at its maximum depth because it will still be growing. It will be growing upwards into a zone where the upper air is divergent and moves outward in the upper part of the developing cell, or moves away from the area immediately above the growing storm cell. These latter conditions in themselves lower the air pressure below them and further strengthen the growth of the storm cell and stretching processes.

Much of this would not be apparent to an observer. The first visual evidence that a waterspout is developing is often a swirling rotation in the boundary layer and some signs of strong wind inflows around it. But observing that requires a perspective from above. To anyone on the water, it may be the changing appearance of the cloud that provides the first warning. At the base of the cloud a swirling column of strong winds may descend from the parent cloud, becoming visible because a funnel cloud forms. The funnel cloud is in the centre of the vortex and is only seen when its falling pressure causes condensation and the resulting water droplets show up the shape of the rotating structure (Figure 6.1a). This may eventually reach the water surface beneath. But it is often the case that before the funnel is seen to descend there is already a swirling disturbance with considerable spray on the water surface. This indicates that the vortex itself has already reached the surface (Figure 6.1b). But it also means the pressure drop within its lower portion has been insufficient for significant condensation of its water vapour into droplets. That may follow if the spin intensifies and the pressure drops further. The waterspout can then be seen in its complete form, although

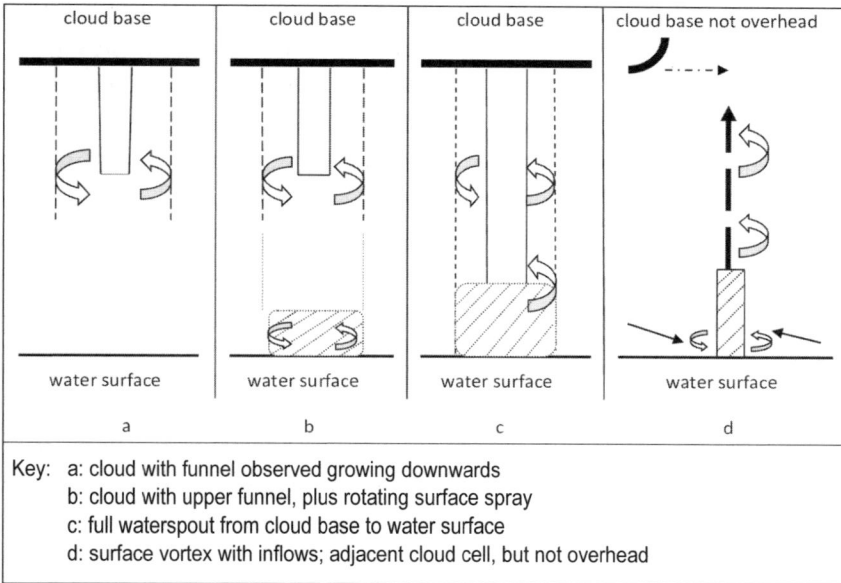

cloud base	cloud base	cloud base	cloud base not overhead
water surface	water surface	water surface	water surface
a	b	c	d

Key: a: cloud with funnel observed growing downwards
b: cloud with upper funnel, plus rotating surface spray
c: full waterspout from cloud base to water surface
d: surface vortex with inflows; adjacent cloud cell, but not overhead

6.1 The characteristics of a waterspout as may be seen by an observer.

normally it would have been present before that happened (Figure 6.1c). For anyone on the water this sequence is a particular hazard. It means that the waterspout can be almost upon them before they can see it!

A frequent variation occurs when a vortex forms in the surface wind flow which is then stretched upwards when a suitably developing cell passes overhead. This may then result in continuing vertical growth until its dynamics are incorporated into the storm cell. At this stage the storm cell is usually growing upwards as well. The whole process is highly dependent upon the convenient co-location of processes at the surface and in the atmosphere above, which is unlikely to occur very often. Thus, there is considerable potential for such events, but a low probability that they will happen (Figure 6.1d).

So far, this demonstrates only part of the dynamic structure of the waterspout. Within the waterspout air is moving in both directions vertically.[19] Golden has shown the outer walls spin upwards carrying water droplets much of the way to the cloud above. This does not include the wider surface spray zone that may be as wide as, or wider than, the waterspout itself. Although the water droplets will rotate with the funnel they will not lift all the way to the cloud base. Golden has further demonstrated that in contrast to this movement, in the central core of the waterspout the air gradually descends as the pressure falls.[20] This accounts for some of the experiences reported by those few who have passed through a waterspout. They report a sense of downward pressure, or of slackening winds and possibly clearer air, in the centre. This downward force, however, can clearly be a significant hazard in itself, as happened in the case of the Roundstone waterspout, County Galway.

The Roundstone waterspout

Each year a regatta is held in Letterard Bay, a small bay on the northern coastline of Galway Bay.[21] Among those who sailed to the event on 10 July 1983 were two sailing enthusiasts from the village of Roundstone, located in the neighbouring bay. There were some heavy showers around when they left, but the forecast was favourable for a good afternoon of sailing.

The two sailors from the village of Roundstone sailed round the nearby headland to the next bay and were amazed at the sight of a tornado funnel sweeping off the land, then coming across the water towards them. As it hit the water surface and appeared to draw up the sea surface water, they could see the full extent of the waterspout as it headed directly towards them. They were too far away from any shelter and had little time to take evasive action before they were engulfed by the swirling wind and water. All they could do was to try to hang on. A flying wet mist enveloped everything, the wind tearing at the boat. In some accounts of similar experiences sailors have spoken of sails and lines getting shredded and torn off, and masts thrust into the water and held there while the remaining outer wall of the vortex passes over. But more significant for the crew are the beams, masts, sheets of sail and any loose objects not secured in time, all flying around in the water sweeping across the boat: injury from flying debris is highly likely at this stage.

But little of this happened in Galway Bay that day. The swirling wall proved not to be the major hazard. The single survivor recalls how, after the boat was hit, it was pushed downwards under the water by what seemed to be a great weight. Both boat and crew were submerged. It was the powerful downward movement in the central core that pushed both boat and crew underneath the sea. By the time the single survivor struggled to the surface the waterspout had moved on further across the bay. But he could not find his friend and crewmate. He had died from a heart attack and would have known little of what happened. This is one of the few recorded fatalities in Ireland from either a waterspout or a tornado.

Waterspouts and Tornadoes

Whether there are fundamental differences between waterspouts and tornadoes has been examined and debated by storm researchers from time to time. Historically, this disconnect has been due partly to the geography of the USA and the different priorities constraining research work there. Early tornado research was focused upon the Midwest, but the primary regions for waterspout research were the Florida coast and the Great Lakes. Both of the latter are about 1500 km away from the Midwest. One result of this was that for many years there was only limited interaction between tornado and waterspout researchers.[22]

The difference between waterspouts and tornadoes

In reality, there is very little difference between a waterspout and a tornado. References to fair weather waterspouts have entered into the literature to try to emphasise what some see as a distinction between a tornadic waterspout and other waterspouts. The former is described as being tornadoes over water, but the latter are said to be different in how they form. This distinction is one that has a significant history in the USA. Golden has argued that waterspouts and tornadoes differ by virtue of their size, intensity, synoptic environment and the dominant role of surface triggers to their formation.[23]

But these contrasts are hard to sustain in a consistent manner, especially when trying to apply them to other parts of the world. Even in the USA waterspouts' size and intensity vary little from the majority of tornadoes that occur on land, for they are also generally relatively small and weak. For many years the emphasis in the scientific literature of the USA was on large tornadoes because of their destructive power. But these were neither representative nor typical of the size and intensity profiles of tornadoes as a whole, even in the USA.

The synoptic conditions that are considered distinctive of tornadoes as opposed to waterspouts are the dominance of wind shear and, in the USA, extremely high CAPE values through the troposphere. In contrast, conditions for a waterspout in the USA are generally considered to be surface convergence, strong instability at a low level and the updraft from a parent cumuliform cloud. Both of these occur in a relatively slack pressure gradient environment with little vertical shear. However, these latter conditions are often part of the tornado-producing environment of land-based vortices and are by no means unique to vortices over water. Finally, the dominant role of surface-based triggers is said to be characteristic of waterspouts when comparing them to tornadoes. However, these same triggers are also found in tornadoes in many environments. So, on the basis of process alone many researchers argue that a waterspout cannot be differentiated in any fundamental way from a tornado.

In practice it is often not possible to make any of these distinctions. Waterspouts appear over ocean and lake areas when they are forming and are reported by eyewitnesses as waterspouts because they are over water. Eyewitnesses cannot be expected to make distinctions between the possible processes that may have produced the vortex. Rarely could any report indicate whether a waterspout has been formed from one set of environmental circumstances or another. So, a waterspout continues to be regarded by many as a tornado over water. Any other distinction cannot be applied with a reliable consistency at present.

An unexpected waterspout off Clew Bay

Some waterspouts occur in seemingly benign weather conditions. Nevertheless they will still have ingredients one would expect of a tornado over land. Such a case was a waterspout observed off the west coast of Ireland on 17 August 2000. It was observed and photographed by a patrol of the Maritime Squadron of the Irish Air Corps during a routine flight, some 111 km off Achill Island, County Mayo, flying below cloud height of approximately 600 m. What was recorded was not dissimilar to other eyewitness observations of waterspouts in Ireland.

It is striking that there was little adverse weather. The only precipitation was well out of sight, so the visibility was excellent at the time. This is characteristic of many waterspout events in Ireland. The

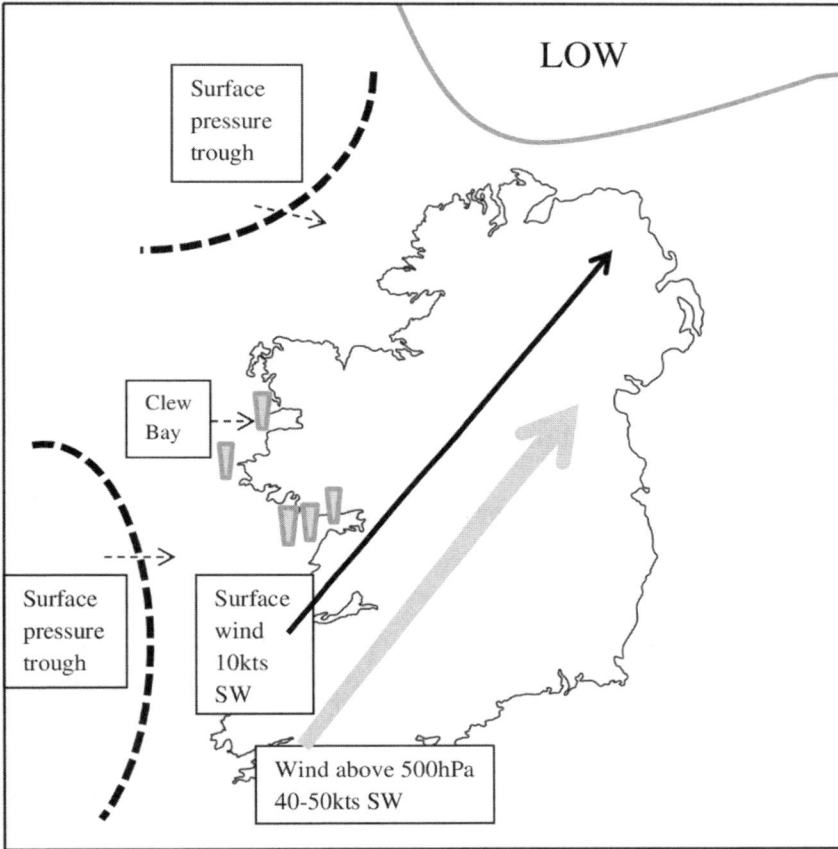

6.2 Synoptic characteristics for the Clew Bay and other waterspouts, 17 August 2000.

vertical soundings through the atmosphere over Ireland at noon showed that winds of up to 10 knots at the surface were overlain by a much faster airflow of 40 to 55 knots at an altitude of 5500 m upwards. This gradient had the potential for producing localised horizontal vortices at this level, just waiting for a mechanism to turn them into the vertical.

Even though generally the weather in the wider region was described as unremarkable with only an odd shower, the parent cloud was a developing cumulus congestus. There were no thunderstorms in the area; sea temperatures were not higher than normal and the winds over the preceding days would have ensured reasonable water mixing that would have spread any localised warming downwards. But there was a surface-related process that did contribute significantly to this event.

The synoptic charts show that the pressure gradient was relatively slack. However, embedded in the wind field at the surface were a series of convergence zones, one of which occurred off the north-west and west coast of Ireland. There was also a limited amount of CAPE concentrated through this lower layer, which would have strengthened the upward effects of convergence and contributed to an upward thrust needed to reorientate the horizontal vorticity into the vertical. This was just one of five waterspouts confirmed along the west and north-west coast of Ireland that day (Figure 6.2). So, a combination of surface convergence in the airflow over the bay, low-level positive CAPE values in the region and wind speed shear between the surface and the middle troposphere provided the basic conditions for the waterspout.

A forecast experiment and the Bravo Platform waterspout

More conspicuously favourable conditions for waterspout and tornado development were present over south-west Ireland on Thursday, 21 October 2004 in developing storms of strong winds, intense rain and hail. The anticipated conditions were such that an experimental forecast was issued by TORRO and published in-house on their website at 06.10 UTC for a possible tornado or waterspout event. The forecast area was roughly the area of Munster and the waters off its south coast. By that time there was already thunderstorm activity near the Kinsale Head gas field, 50 km off the Cork coast.

At 07.35 UTC a waterspout, partially hidden by a rain and hail shaft, was seen approaching the gas extraction platform Bravo by the pilot of a 4-tonne Eurocopter SA365N, transporting passengers from platform Bravo and its partner platform Alpha, a short distance away. The waterspout was about 1 nm off Bravo and heading towards it. Photos show a vertical shaft between the cloud base of the storm cell and the sea surface. So, keeping the rotors running for rapid take-off, the pilot urged the passengers off the helicopter to clear the deck for a rapid departure.

The final report of the subsequent enquiry recorded that the helicopter was battered by the intense wind, rain and hail and that it rocked and shook violently. As it was engulfed by the vortex the helicopter rolled and tilted first one way and then another as large horizontal changes in wind direction occurred, as well as strong vertical motions. The helicopter rolled and yawed, its nose lifted and then plunged before lifting again. 'With only limited visual clues available due to the intensity of the rain and hail, the pilot fought to regain control of the helicopter, while in close proximity to the deck'.[24]

The enquiry concluded that the incident was due to a gust front or downdraft from a thunderstorm and did not appear to pursue the possibility of a waterspout. It was also quite unaware of the TORRO forecast. However, the pattern of the winds and the behaviour of the sea do not appear to match a downdraft. The large swings in wind direction as well as the strong vertical motions referred to in the report would be very unlikely in a downdraft, when winds fan out from a downward flow. In contrast, wind direction in a rotating vortex fluctuates violently. Also, a downdraft surges forward at the surface as the rushing downward flow of air meets the surface and spreads out, fan-like. The photographs that record the event show no forward tongue-like motion at the leading edge of what was described as a squall. Indeed, the photographs show that the rain-wrapped waterspout had sharp vertical edges right to the sea surface and a classic wall cloud above it extending downwards beneath the main storm cloud.[25] Platform Alpha had a meteorological station with a full range of instruments. During the event these recorded significant changes in atmospheric pressure. But the data was dismissed in the subsequent enquiry as probably the result of a temporary malfunction. So, the eyewitness observations, the photographic evidence and the meteorological data are all consistent with this being a waterspout.

Further around the Munster coastline the upper air soundings taken at Valentia meteorological station both before and after the event showed conditions were conducive for such an event. Indeed, this is why TORRO issued their forecast in the first place. At midnight prior to the event there was a marked shearing of the wind speed and direction which became much stronger at a lower level as the morning progressed. At the same time the sea area off the south-west coast became overlain by the left-exit region of the jet streak in the upper westerlies from the Atlantic. This would have facilitated vertical stretching, as would the convergence of airstreams at the surface which had contributed to what the enquiry's report described as 'a very active showery trough [was] moving through the area.'

The result of this was that the experimental forecast was considered to be a success, even though the forecast was not produced from Ireland itself. It provided a basis for continuing with further such forecasts to cover the whole of Ireland as well as the UK, where it originated from.

Inland waterspouts: Lough Neagh events

Lough Neagh is by far the largest of Ireland's inland loughs, many of which have a long history of waterspouts. The lough is very visible from many miles distant from its shore, since not only has it the largest freshwater surface of any lake in Britain or Ireland, but it is also set in a wide basin with a high rim of mountains, often over 400 m high. As a result, there have been many eyewitnesses to these events and, in more recent decades, the photographic record held by the local population has grown steadily.

Such an event occurred on 3 June 2014, very suddenly to the surprise of seasoned storm watchers. However, a number of photographic images successfully captured the waterspout from varying distances across the basin, including one taken from a mobile phone (Figure 6.3). That this waterspout was a surprise was due to an extensive cloudscape that exhibited widespread capping that limited vertical cloud growth. But within all that, a highly localised interaction of a group of processes (such as surface convergence, sharp vertical differences in wind speed and/or wind direction, energy release from evaporating water droplets and localised solar heating effects) generated the spout very suddenly. This emphasises how very unpromising skies may conceal significant localised processes that can produce dramatic weather effects, because we can neither see nor know everything that is going on above our heads at any one time.

The history of waterspouts on Lough Neagh is centuries old. Indeed, the very name of the lough is linked to the phenomenon. The name refers to the horse god Eochu, who is said to dwell under the water. From time to time, fishermen and others on and around the lake have heard what they described as booming noises from beneath the water. This is said to be Eochu galloping to the underworld. Over the ages, the timing of this phenomenon has been directly associated with waterspouts and

6.3 The waterspout that developed suddenly on Lough Neagh, June 2014. (Photo by Gemma Convery)

whirlwinds of various kinds and such references remain in the region's folklore. The lough is now legendary for its vortex activity.

Eddy Waterspouts

There are cases where a surface vortex results in a short-lived spin-up, when a rotating spout may whip up the water surface quite violently, in a vortex that may grow as tall as 25 to 50 m. They travel short distances but still pose significant hazards to water users. These phenomena are eddy whirlwinds or, more appropriately, eddy waterspouts. They form quite differently from the waterspout vortex that stretches between the water and an associated cloud. In Ireland, they are a product of both strong winds and the surface terrain. Such eddies are created once the wind speed reaches a critical intensity, depending on the detailed geography of the terrain.

In coastal Ireland they are associated with relatively low spits of sand hills enclosing coastal bays (as in Tralee Bay, County Kerry), promontories with cliffs that extend seawards (as near Mizen Head, County Cork) or large conical hills on promontories that strong winds sweep around to cross expanses of water on the leeward side (as in Blacksod Bay and Achill Sound, County Mayo). Inland, the lakes of steep-sided glaciated valleys, especially in the west and north of Ireland, are particularly associated with such waterspouts. From time to time local fishermen and other water users have to calculate the risks of these hazards when on these waters. Because of their relative remoteness and the sparseness of population, there has been little general awareness that these events are part of the long-term environmental profile of these localities. But there are a few exceptions.

The lakes of Killarney

Killarney became a tourist destination from at least 1747 onwards. As it opened up, so the school of topographical writers that blossomed in the Victorian era gave it some attention. One of these, Dublin-born Isaac Weld, includes in his account of the lakes at Killarney:

> A gentleman living near Killarney, who had often crossed the ocean, assured me he had more than once beheld it so much agitated by the impetuous hurricanes which descend in circling eddies through the passes between the mountains, that its waves, drifted together and raised to an immense height above the surface, have assumed the terrific aspect of a water-spout. Though such tremendous storms are seldom experienced in summer, yet as squalls occur even during that season, no boats should be used that are not able to encounter heavy waves.[26]

These events have been given the local term 'rowdow' by those familiar with them, particularly those who have been fishing on the lakes for many years. The term is thought to have been introduced during the nineteenth-century tourist boom. Most rowdows occur on the Upper Lake and the Middle Lake, although some do occur on the Lower Lake as well. The direction of the wind is significant as to where they occur. Southerly winds may give rise to rowdows on the Lower Lake from the Dinis Peninsula, but the stronger ones come down the mountain from Benson's Point. A strong east wind, however, will produce them in the Middle Lake from Torc Mountain into the bay west of Brickeen Bridge. Rowdows are always associated with the arrival of particular stormy conditions, a falling barometer and sheet rain (as it was described locally). Initially they tend to be about 5 m across. The rowdows grow vertically as they cross the lake, their spray rising to 100 m or so. Sometimes they occur in frequent succession. On encountering these, local fishermen need 'a very cool nerve, experience and knowledge of what their boat can handle, in order to avoid capsizing, since in the stormy conditions you have only about thirty degrees of leeway to point the boat using an engine at very low revs or oars when a rowdow is coming at you. A rope out back gives stability.'[27] When caught, boats have been spun around, several times, depending on the diameter of the rowdow, so damage tends to be minor although its potential is much greater for the inexperienced.

Doo Lough, County Mayo

Within Connemara, the lough with the strongest reputation for eddy waterspouts is in the uninhabited Doo Lough Valley. With only boglands and steep mountainsides all around it, there is no resident population, and it is only the adventurous, travelling through this narrow towering pass alongside the dark peaty water, that have reported these events. But that has been often enough over the years for them to become known as 'white ladies', although the term 'water devil' is also used (Figure 6.4).

The lough is aligned north-west–south-east over a relatively short distance, being about 3 km long and up to 0.5 km wide. But when storm winds are appropriately aligned, they accelerate through the valley and

6.4 The 'white lady' of Doo Lough. (Photo by Roger Derham)

create waves large enough to break onto the narrow mountain road, posing a hazard to local people using the pass. The eddy waterspouts often form at the southern end of the lough. They may occur when a turbulent, strong wind from a southerly direction accelerates and becomes more turbulent as the valley quickly reaches its narrowest. The vortices that form can be erratic in their movement, sometimes being stationary and at other times accelerating. They are a threat to anyone on the lough as well as along the shoreline. Local traditional folklore has references to them and their effects.

Achill Sound, County Mayo

Because eddy waterspouts are produced from a combination of wind conditions and the local terrain, it happens from time to time that successive eddy waterspouts occur while the conditions persist. Such a family of eddy waterspouts was witnessed near Achill Island on 26 October 2005, another location where there is a long history of such events.[28] The strong southerly winds at the time had to pass round the conical hills of Corraun, overlooking Tonregee and Achill Sound. The upper air soundings at both Valentia (to the south) and Castor Bay (to the north-east) recorded an intense wind speed gradient in the lowest part of the atmosphere, being 10 to 15 knots at the surface and a dramatic

40 to 50 knots at 1300 m. Such strong wind shear had a very high potential for the development of vortices, being further enhanced by the irregular surface topography.

From the shoreline of Achill Sound, an extension of Blacksod Bay, the vortices were observed and recorded as they arrived and hit the water surface, with a roar like a large truck travelling at speed on a wet surface. It was late in the day, but still the spray revealed the vortices from about 50 m offshore, with a diameter of 20 to 25 m. Each was timed and plotted as it crossed the bay, the longest tracking to the opposite shore at Inishagoo, some 4.75 km distant. Even in the fading afternoon light their progress could be monitored by their spray, which disappeared when they moved off the water on reaching the opposite shore. Between 16.15 UTC and 16.40 UTC, five of the twenty that were recorded followed this longer route, while the others deviated to the islands of Inishbiggle and Annagh, on either side of the main channel. The distance of 4.75 km was covered in times that varied between 35 to 42 seconds. That is a staggering speed of 336 km/h to 480 km/h! The meteorological station at Belmullet, at the northern end of Blacksod Bay, recorded surface winds of 27 knots, gusting to 41 knots at the same time. It seems unlikely that many fully developed waterspouts travel as fast as this, although it is normal to underestimate speed over water because of the lack of reference points.

These observations highlight the hazardous nature of such events. The poor visibility at dusk can make their approach in the half-light very difficult to discern. But the speed with which these half visible vortices can travel, as on this occasion, makes them a very serious threat to anyone in their vicinity.

Similar events occur in the same locality when strong winds arrive from the west. On the western side of Corraun there have also been many cases of eddy waterspouts, described as stretching vertically to 50 m or more, crossing Bellacragher Bay, an extension of Blacksod Bay, to its south-west. Locally, they are known to have a particular track across the bay. This reflects the particular alignment of the mountains with respect to the wind direction and wind speed required in order to produce the vortices. Across Bellacragher Bay they reach the eastern shoreline after approximately 500 m.

Such events are well known by local inhabitants around Blacksod Bay and its extensions. Their long, though spasmodic, history has resulted in a traditional awareness that goes back generations. As with many parts of Ireland, the relative remoteness and sparseness of the population results in the knowledge of these events remaining local most of the time.

Donegal mountain loughs

There are a number of loughs in County Donegal in the vicinity of Mount Errigal that are known for their waterspouts. One of these is Lough Beagh, located in the Glenveagh National Park, whose website is one of the few in Ireland that explicitly warns of dangers from waterspouts. There they occur in stormy weather when there are near gale force winds in the lough area. October onwards and the rest of the winter months are most favoured. These waterspouts can be visible to a height of 40 to 50 m, but dissipate as soon as the shoreline is reached. It is estimated by the park rangers that four to five eddy waterspouts occur per year. Lough Beagh is a very narrow lough, about 6.5 km long and 1 km wide, orientated south-west–north-east with tall steep sides. The south-west half of the lough (the Owenbeagh valley) has almost cliff-like sides reaching to 396 m which extend as a long, almost dead-straight gorge from the higher ground to the south-west for 3 to 5 kms before widening slightly into the lough. Here, downstream vortices briefly form in the windflow from time to time. When strong upper winds pass across the summits above the valley, a 45-degree wind shear can be created with strong winds below, setting up an environment in which the vortices can be stretched and waterspouts generated on the lough. These invariably travel from south-west to north-east.

In the neighbouring valley, at the foot of Mount Errigal, there are two more loughs that experience waterspouts: Dunlewey Lough and Altan Lough. However, the normal direction of movement is quite different from those of Lough Beagh. Sometimes strong winds from the east or north-east occur that accelerate around Errigal Mountain. This is known locally in Dunlewey village as the 'Errigal wind'. Occasionally vortices spin off as the Calabber valley descends onto Dunlewey Lough on the south side of Errigal. These vortices are stretched as they descend to the lough and accelerate, completing their development. Eyewitnesses have described them as 'reaching the clouds above' and sometimes tracking across the lower western shore where they have caused damage, continuing onwards as weak tornadoes. They occur at any time of the year, winter or summer, providing the wind conditions are suitable.

On the north side of Errigal mountain is Altan Lough. It is reported that very occasionally a similar set of circumstances occur there as at Dunlewey Lough. Altan Lough is within a valley that descends sharply towards the north-west and north, initially severely confined between the cliff-like slopes of Errigal and Aghla, which flank it on either side. So waterspouts here generally track northward.

Doagh Isle, Inishowen, County Donegal

Some local people regard two to three waterspouts a year as being the normal event frequency for this small area. The Isle of Doagh faces westwards to the Atlantic as it approaches Lough Swilly to the south. It has very steep-sided mountains on its southern side, with a series of peaks rising to between 250 m for Mount Binnion, right on the coast, and 502 m for Raghtin More. Between these peaks are some extremely narrow valleys that descend rapidly towards the Isle of Doagh and Trawbreaga Bay. When very strong winds sweep through these valleys, vortices may form offshore as the funnelled winds converge, aided by a highly irregular coastline which creates sharp wind gradients in speed and direction at the surface. Some of these vortices grow to a few tens of metres, but occasional ones are stretched by passing storm cells into which they are drawn. There are a number of ways in which such convergence can be caused in this setting, but very few occasions when the conditions all come together to produce either the eddy waterspout or a full waterspout. It is likely that the frequency of two or three per year is an underestimate because there will be no one with an awareness of all the events that occur. Once the waterspouts have formed and moved across the water, they sometimes cross the shoreline. They have acquired a history of damaging impacts which includes structural damage to houses, the dislodging and overturning of farm machinery and the scattering of large stacks of peat.

Carlingford Lough, County Down

In Carlingford Lough, County Down, eddy waterspouts form and are largely confined within the lough. The lough has a length of 16.5 km and widens to a maximum width of 5.5 km. The mountains on either side funnel the winds from particular directions, so that they accelerate downwards to the lough. Strong west-north-west winds from the hills near Rostrevor produce many of the vortices, although others are caused by strong south-south-west winds that pass over Carlingford Mountain. The vortices are formed initially as lee eddies in both cases. These stretch downwards as the terrain falls away sharply. Limited case studies suggest that this is often beneath a strong low-level inversion that inhibits vertical upward stretching but provides no barrier to downward stretching facilitated by the terrain. As in Blacksod Bay, waterspouts may occur in clusters, one forming after the other, then moving away downwind.

Locally, these waterspouts are known as 'kettles'. This term originates from the Old Norse *ketill*, which means cauldron. This term captures that swirling confusion of air and water droplets that appear to mimic boiling water. The *Irish Annals* indicate that the Vikings were establishing themselves in the Carlingford Lough area from the ninth and tenth centuries AD, so the eddy waterspouts have a history extending well over a thousand years: there is nothing new about them.

Both waterspouts and eddy waterspouts occur frequently in numerous locations. Throughout Ireland stories of these phenomena abound. But modern decades have seen the population of the remoter lakes and coastal regions decline, so that the awareness of these phenomena has declined significantly as well. Today, those familiar with them rarely report them and do not consider they require anything other than a passing comment.

CHAPTER SEVEN

Tornado Strikes in Ireland: Chronologies and patterns

*There will always be a bigger tornado coming
and it may well be the very next one.*

T
hese words of advice have been given to trainee forecasters in
the USA. Very infrequent weather events tend to enter the
remote world of statistical probability where any sense of a
possible imminent calamity is easily lost. But expectation is the key to
anticipation. Scientists find very low frequency events are very hard to
handle because not only are they inadequately represented in weather
databases, such events would also be outside the experience of most
people. In Ireland, tornadoes are one of several such infrequent weather
types, especially with reference to particular localities and regions across
its landmass.

Tornadoes are often associated with particular places. In any
discussion references to the American Midwest and its 'Tornado Alley'
quickly surface. Early films about tornadoes that fascinated international
audiences were set there. Thus *Twister* (1996), *Tornado Alley* (2011) and
several others reinforced the association.[1] For many people a tornado
poses a threat, and they are concerned at the level of risk they may be
exposed to, even here in Ireland. In order to normalise and contain the
potential tornado threat, it seems that people would like to know where
they are most likely to occur. Once that is known, there is a sense that
the risk they pose is calculable and, seemingly, manageable. It is often

111

thought desirable that the process of scientific study should lead to such conclusions; not merely to an understanding of how and why tornadoes occur, but where, when and how frequently they strike. Other people are interested in tornadoes primarily because of their curiosity value, to satisfy a legitimate desire to better understand the world we live in. For both of these reasons, one of the first steps in studying this phenomenon is to build up a database of events. From this it would be possible to extract patterns that can help in responding to these demands.

Irish data have now been carefully recorded for many years. For the nearly twenty years between 1998 and 2017 all known reports from within Ireland have been investigated and verified by site investigations. This rigorous on-site verification procedure has made the Irish data set a particularly comprehensive and reliable one for that period, comparable at least to any tornado data set available. Most others are based largely on reports alone, with only an extremely limited number of cases examined in detail by an on-site verification. The Irish work has been based at University College Cork and supported by TORRO, a research organisation that has done much to stimulate tornado research in Europe.[2] Earlier events have also come to light, both of recent decades (pre-1998) and, as has been demonstrated, of the more remote historical past. So the record has been expanded and modified. The creation of this data set has been a lengthy, but important, task. It is the purpose of this chapter to assess the state of our knowledge to date from the detailed studies of the last two decades, with some additional reference to the more limited information from earlier years.

Annual Patterns

Tornado totals

In the twenty years of detailed study, a total of 189 tornadoes (including waterspouts) have been recorded and verified for Ireland.[3] These have been carefully distinguished from funnel clouds and other vortices. Although twenty years is a relatively short period, these data mean that approximately nine tornadoes, in addition to the many funnel clouds that did not develop into tornadoes, have occurred each year. Of course, they are not spread evenly between the years; the annual totals fluctuate considerably. However, these data do help us to gain an insight into the reality of tornadoes and associated vortices occurring regularly within Ireland.

During these twenty years the reports of tornadoes have been systematically located, verified and recorded. Although all known events have been included in this process, a small number may have been missed. Before 1998, only a small proportion of individual cases reported had been investigated on site. Most of these earlier tornadoes have been entered into a longer database that has also been built up. These have been dependent on a wide variety of other sources, each of which has been evaluated and checked. Some of these investigations have brought to light events in different parts of the country that were not

Table 7.1 Annual tornado (TN) data, 1998–2017.

Annual Tornado Profiles					
Year	Annual Total	5 Highest (•) 5 Lowest (○)	TN days	TNs per TN day Lowest → Highest	T value range Lowest → Highest
1998	7		5	1 → 3	T_2 → T_4
1999	14	•	12	1 → 3	T_0 → T_5
2000	15	•	9	1 → 4	T_0 → T_3
2001	13	•	9	1 → 4	T_2 → T_4
2002	12		10	1 → 3	T_0 → T_4
2003	6	○	6	1 → 1	T_0 → T_3
2004	12		12	1 → 1	T_0 → T_2
2005	12		6	1 → 6	T_2 → T_3
2006	15	•	10	1 → 2	T_0 → T_4
2007	15	•	14	1 → 2	T_0 → T_2
2008	9		9	1 → 1	T_0 → T_2
2009	5	○	5	1 → 1	T_0 → T_2
2010	7		7	1 → 1	T_0 → T_1
2011	5	○	5	1 → 1	T_1 → T_3
2012	7		6	1 → 2	T_1 → T_4
2013	8		4	1 → 3	T_0 → T_3
2014	10		7	1 → 3	T_1 → T_3
2015	2	○	2	1 → 1	T_1 → T_1
2016	5	○	5	1 → 1	T_0 → T_2
2017	7		6	1 → 2	T_0 → T_2

widely reported, or even reported at all, when they originally occurred. Therefore, the data for the pre-1998 period cannot be considered to be complete in any way. But the data for the period after 1998 can be regarded as a reliable, comprehensive record, within the constraints that are discussed in this chapter.

The annual totals of tornadoes and waterspouts that occurred during this twenty-year period are quite variable. Although the average is nine events per year (actually 9.3), Table 7.1 shows that the last twenty years have dived between those that have been relatively active and others that have been less active. The active years had twelve to fifteen events, while the contrasting less active years had two to seven events. The most common annual total was seven, but the second most common were fifteen, twelve and five (all equally frequent). So, the average year has only occurred once! Thus, the two ends of the frequency spectrum have tended to cluster. Most of the last few years have been among the less active ones. So it is more meaningful not to refer to a single figure when referencing the Irish experience of tornadoes and waterspouts. Instead, the record shows that there are normally between five and fifteen events each year.

Tornado days

It is possible that a more useful measure of tornado activity and its potential threat to Ireland is the number of tornado days rather than the actual number of individual tornadoes. In particular, this may assist in building up a greater public understanding of the relatively unfamiliar hazard to which it is exposed. In a small country like Ireland, everywhere is vulnerable at some time or other during the year, even though only a very small area is actually impacted by any one tornado. If effective tornado forecasting to pinpoint high-risk days is developed, it could then be matched by a relatively informed public awareness and an effective response.

On average, there have been just seven tornado days per year (the data in Table 7.1 give a precise value of 7.5). But the difference between years is considerable. This ranges from the quietest year with only two tornado days (2015), to the most active year when there were fourteen tornado days (2007). These and other data in Table 7.1 show (again) how the average is a poor indication of what to expect.

A significant proportion of the days on which tornadoes have occurred have registered only a single tornado. Indeed, this has been the case for eight of the twenty years. But there have been as many as six on a single day, as on 1 January 2005. In that case the tornadoes were spread over four counties, from County Westmeath to County Armagh. Such an occurrence is relatively unusual, but it has not been so unusual for individual days to have up to three or four tornadoes in different locations during a year.

One of the important characteristics of a tornado is its intensity, or the speed of its inner winds, because this largely determines the damage it can inflict. The range of tornado intensities for each of the twenty years is given in Table 7.1. Strong and severe tornadoes (T3 and T4 respectively), or even more intense events, are particularly alarming. Eight years of the first decade recorded here had one or more such strong or severe tornadoes or greater, five of them being in the upper end of that range. The second decade had fewer destructive experiences, only four years showing events within this range and only a single year reaching the T4 intensity. But making such distinctions should not be at the expense of undermining the severity of any of these events: even a T0 tornado may have serious consequences.

The annual variability in tornado-related phenomena shows that although there have been similarities between years there is no individual year which can be said to have been typical. The annual patterns are not simple, even though the range of values is not large. The first of the two decades show a slightly higher frequency in tornadoes and tornado days as well as slightly more intense events. But how significant that may be remains to be seen. This is also striking in the context of a significant increase in the enthusiastic weather-observing community across the country. In addition, there has been an ongoing development of means for reporting such events that have now become so much easier to record. Often it has been assumed that this leads to a noticeable increase in localised events, such as tornadoes, being reported. Clearly, the Irish data shows that this is not necessarily the case.

Regional contrasts

The regional variation in the number of individual tornadoes that have occurred across the country during this period has been considerable. An assessment of their occurrence by county demonstrates some apparent geographical preferences. Over the twenty years up to 2017, crude frequencies show that the three largest counties of County Cork (8), County Mayo (8) and County Galway (6) had more tornadoes than most. However, some much smaller counties surpass these totals, the three most significant being County Westmeath (12), County Antrim (10) and County Wicklow (9). But the size of each county is always going to distort such comparisons when the size difference between them is large. So, when the tornado numbers are adjusted by the area of each county, an indicator of tornado frequency that has occurred per 1000 km² tells an interesting story. This more realistic comparison shows that the most affected county during the last twenty years has been County Westmeath, in the midlands, followed by County Louth and County Wicklow in the east. At the other end of the scale, the smallest number of verified events took place in County Tipperary, followed by the adjacent counties of Kilkenny and Clare. The results are shown in Figure 7.1. In this case they exclude all offshore waterspouts since many of them cannot be allocated to the offshore waters of an individual county with certainty.

It is striking that there is no county in Ireland that has not experienced a tornado at some time or other during these twenty years. Indeed, some counties would have been buffeted by more tornadoes than these figures state. This is because the data identifies where the tornadoes made their initial contact with the ground surface. They then tracked across the countryside for distances that may have taken them into neighbouring counties. For example, in 2006 a tornado tracked for 30 km, starting in County Armagh and continuing with most of its track length in County Antrim. Similarly, in 1995, a tornado tracked 29 km from County Westmeath into County Meath. But these are exceptions. Irish long-track tornadoes are few and far between, so not many tornadoes will cross over from one county to another.

Although the concept of a tornado alley is vague and limited in value, it is so often embedded in any discussion of tornadoes that it is

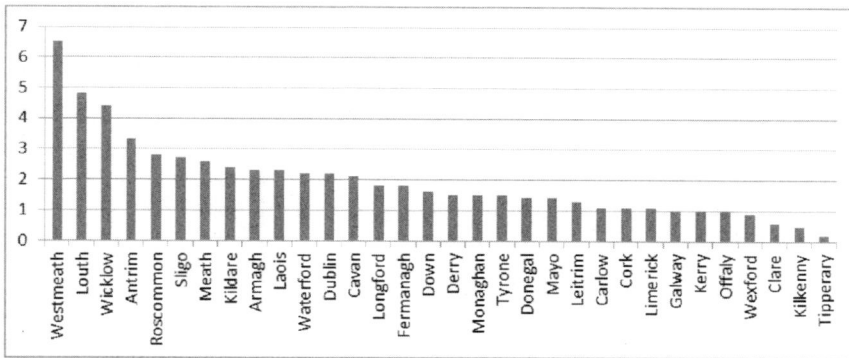

7.1 Annual tornadoes and waterspouts per 1000 km², by county, 1998–2017.

inevitable for the question to arise as to whether there could be such a phenomenon in Ireland. The distribution of events shown in Figure 7.2 is such that it is tempting to suggest Ireland has a tornado alley across part of the midlands. But it is necessary to emphasise that this term is only a colloquial nickname, without any scientific definition. It is mostly associated with the popular media, as well as being a conversational convenience, to refer to a continuous swathe of territory with relatively high tornado occurrences. Its use tends to place a disproportionate focus on a limited area, thereby overlooking those areas far beyond this zone that are also at risk. In Ireland, the relative proximity of the band of counties with the highest tornado frequencies may make this concept an attractive idea. Indeed, it is reinforced by the counties with the lowest frequency being all along the southern edge of this zone, creating a relatively sharp boundary when mapped in county units. But a much longer and larger data set of events will be required before the term can be used for Ireland with a longer-term certainty of its geographical validity.

Over the twenty years up to 2017 the number of tornadoes and waterspouts in any one county during an individual year has been found to be relatively low. No county gets a tornado every year. On the other hand, some have been on the receiving end of more than one in a single year, but that does not happen often. Indeed, during this twenty-year period the highest number of years that any county has suffered a tornado has been eight. This unfortunate distinction is shared by County Donegal and County Antrim. The latter includes a large proportion of

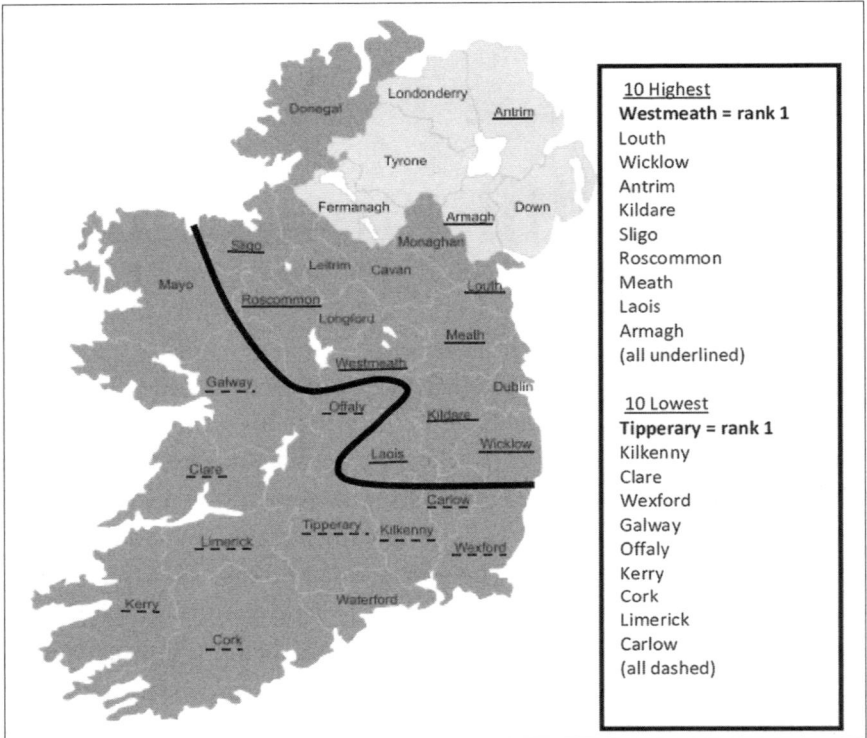

10 Highest
Westmeath = rank 1
Louth
Wicklow
Antrim
Kildare
Sligo
Roscommon
Meath
Laois
Armagh
(all underlined)

10 Lowest
Tipperary = rank 1
Kilkenny
Clare
Wexford
Galway
Offaly
Kerry
Cork
Limerick
Carlow
(all dashed)

7.2 The geographical divide between the ten counties with the highest number of tornadoes per 1000 km² (solid underlined) and the ten with the lowest number of tornadoes per 1000 km² (dashed underlined), 1998–2017. The individual highest and lowest are in bold.

Lough Neagh where some of the events were waterspouts. But again, the contrasting geographical size of counties would be a significant factor in any full comparison. Many other counties are close behind, each with five or more tornado years.

Reference has been made to the remarkable geographical divide between counties with the highest and lowest frequency of tornadoes (per 1000 km²). Of the thirty-two counties, the ten with the highest frequency are all adjacent to one another and, therefore, form a zonal block across the central and eastern regions of the country. Similarly, those with the lowest frequency are also adjacent to each other, forming a group of coastal and inland counties in the south and west of the country. It is also noteworthy that these two groups are themselves immediately adjacent to each other. So, a distinct boundary line can be drawn between them, from County Sligo on the west coast to County Wicklow in the east (Figure 7.2).

Topographic associations

There is no obvious explanation for this remarkably sharp transition between counties with the lowest to those with the highest tornado and waterspout frequencies. But it may well be worth noting that there is a marked contrast between these two regions in terms of altitude and topography. The southern and western counties are generally higher in altitude. Linked with this is their greater topographic diversity with mountain slopes, moorland summits, steep terrain and, often, quite narrow valleys. Such characteristics are much more muted in the central and eastern counties referred to above, although they are present to a limited extent.

This association can be developed further by making a generalised comparison in terms of the highest altitude and then the altitudinal range in each county. The group of ten counties with the lowest frequency of tornadoes and more mountainous terrain have a mean maximum altitude of 739 m. Seven of these ten have summits above 609 m. In contrast, the ten central counties with the highest tornado frequency that are lower lying and with less topographical variation have a lower mean maximum altitude of 569 m. In addition, seven of its ten counties have summits below 609 m. Another measure of this contrast is in the elevation range within an area, rather than its mere altitude. This is a closer reflection of the surface topography. But the result differs very little in this case; the two regions emerge just as distinctly.

Tornado research in other countries has explored the possible role of the surface terrain in the development of tornadoes, particularly with regard to the role of mountain areas.[4] Mountain conditions that have been linked to a comparatively reduced tornado frequency within the mountain zone include mountain air temperatures and the modification of low-level airflow in passing storm systems. In particular, colder mountain air tends to have a greater stability, so one of the contributing factors in tornado development is very weak in these areas. In addition, the mountains may cut off inflow into passing storm cells and thereby repress a potential contribution to vortex development. The more complex the surface terrain the more inhibiting and disruptive it is likely to be to the surface inflow and the lower part of any developing vortex.

The presence of a mountain area has been linked to a relatively greater frequency of tornadoes in adjacent regions beyond the mountains. This is due to two factors. Firstly, the sudden decrease in elevation on the lee side of a mountain may cause an existing updraft to stretch and tighten, thereby rotating faster (vorticity stretching), which then tracks into the area beyond the high ground. Secondly, vortices may be initiated and shed at the end of a strong relief feature. They may act like a rock in a stream, generating wake vortices, having passed by the mountains. A storm may then ingest such a vortex and stretch it to produce a tornado. This also occurs in coastal regions where vortices are shed from peninsulas.

There has developed a relatively widespread agreement that the effect of the surface terrain upon tornado development is greatest when the measured intensity of a tornado is within the lower half of the scale of intensities that are physically possible. Most Irish tornadoes are in this category.

In the case of Ireland, many of the storm systems approach its coastline from the west or south. This onshore air brings relatively cool, maritime air, which lacks any strong instability and is, therefore, a relatively unfavourable environment for frequent tornado development. When the upper air soundings taken at Valentia are related to tornado events, this effect is generally apparent since convective indicators such as CAPE tend to be weak and wind shear and helicity indicators are rarely promising. But tornadoes still occur in and around the mountain areas nearby. The terrain within the mountain areas may itself produce surface vortices in the airflow which, when stretched upwards by high-level cells, may result in a fully developed tornado vortex, even in cool weather conditions. A further situation that produces a tornado within these mountain areas is when an offshore waterspout comes ashore and then tracks across the coastal area and beyond. When this occurs the continued existence of the vortex will be largely determined by how the terrain changes the surface inflow essential to the persistence of the vortex. In most cases this is likely to result in a slowing and deflection of the inflow, so that the vortex dissipates quickly.

There is a lot that goes on in mountain weather as storms pass through and over them. However, we hear little about these details. Indeed, they may not even be observed. One consequence of this may well be that the relatively low frequency of tornadoes in these higher areas may be simply that they are poorly reported compared with other counties. This factor is considered to have been a significant possibility in other countries where high relief is associated with fewer recorded tornado events.[5] In Ireland, the number of inhabitants is relatively low in the mountain county group, which have a population density of 19/km² compared with 36/km² in the midlands and east coast group. That is only 54 per cent as many potential observers as in the latter region. But possibly of greater importance is that the visibility of these events is significantly reduced by the varied relief and resulting higher skyline. However, it is unlikely that these factors make a significant difference to the record.

Seasonal and Monthly Patterns

Monthly frequencies

Monthly totals of tornado events appear to show strong seasonal differences in tornadic activity in Ireland. Table 7.2 shows that tornadoes occur most frequently in the summer months. August events dominate the data overall. Approximately 19 per cent of recorded tornadoes occurred during this month alone. However, this dominance is by no means evident every year. In fact, there were only five years when August was the lead month for having the most tornado events, an honour that has been shared by eight of the twelve months. Thus generalisations regarding any seasonal and monthly dominance for these events may be very misleading. Indeed, there have been years when January (as in 2005) or December (as in 2006) have been the most active. Nevertheless, the months of the summer half-year are dominant overall. The six months on either side of the equinox are April to September, which account for 65.9 per cent of the total. But the actual tornado frequency peak months are May to October, which account for 71 per cent of the total.

Table 7.2 Monthly distribution of tornadoes and funnel cloud characteristics over twenty years.

Monthly Tornado Profile				
	Tornadoes % per month	Funnel Clouds % per month	Single-day TN clusters % per month	Tornado days	Funnel Cloud days
20-year total	**186**	**442**	**48**	**139**	**253**
January	7.0	0.2	4	8	0
February	1.6	0.9	0	2	3
March	4.3	0.7	0	7	3
April	4.8	7.7	9	9	27
May	9.1	12.0	13	17	31
June	10.8	18.8	9	17	45
July	11.3	28.6	18	19	68
August	18.8	24.4	18	25	53
September	10.8	3.4	10	13	12
October	10.2	2.7	9	11	9
November	5.9	0.2	6	4	0
December	5.4	0.5	4	7	2

Winter and summer funnel clouds

This pattern is further reinforced by the behaviour of funnel clouds. These are vortices that have not developed into tornadoes and are much less significant in terms of the risks they pose. Some funnel clouds are notably benign, being small, very transient and frequently associated with a dissipating parent cloud. Often, this type of funnel cloud is not associated with severe weather at all. Most reports investigated in Ireland have rarely been of this type. Instead, the majority have proved to be structures that have even been mistaken for a growing tornado, with an intense columnar vortex extending significantly towards the ground surface. Figure 7.3 shows such an example. These funnel clouds are a sharp contrast to their benign cousins.

Nevertheless, both of these types have a noteworthy role in shaping the hazard potential for any airborne activity when they do occur. In addition, they raise the awareness of the greater hazard posed by a tornado, with which they are sometimes confused in popular reporting.

7.3 A funnel cloud near Maghera on 8 June 2011. (Photo by Martin McKenna)

Table 7.2 shows the percentage of funnel clouds that occurred each month during the twenty years up to 2017, when 442 such events were recorded.

Again, it is striking that for funnel clouds the summer months clearly dominate. There is a much narrower peak for these events than is the case for the peak months for tornadoes. The dominant month is July with nearly 29 per cent of all occurrences. But together, the three months of June, July and August account for 72 per cent of the recorded events. However, it is likely that the record for funnel clouds is much less complete than for tornadoes and waterspouts because they do not impact the ground surface directly. They depend upon observation alone and the readiness of the observer(s) to report them. In particular, the shorter day length of winter months will always affect the record.

Seasons of threat

It is a feature of all the data for tornadoes, waterspouts and funnel clouds that they have occurred in every month of the year, including the winter months. To be content with the general statement that most tornadoes and tornado-related phenomena occur during the summer months would be to make an overgeneralisation that could mislead one into

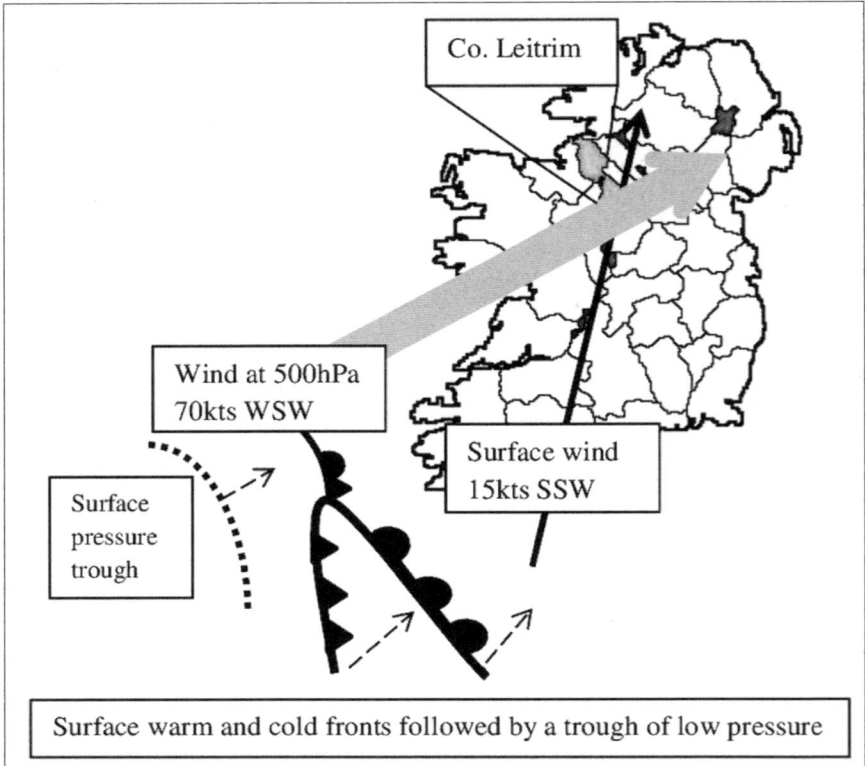

7.4 Synoptic characteristics for the Carrigallen tornado, County Leitrim, 26 January 2002.

overlooking the threat posed at other times of the year. Between the summer and the early winter, the Irish atmosphere is very different and undergoes a number of important seasonal changes. This is supportive of the case that there is not one set of weather conditions that give rise to tornado features over Ireland, but a number of quite varied ones.

The inadequacy of any assertion that there is a summer tornado season in Ireland is reinforced by a consideration of the most serious tornadoes that have occurred. There have been five intense (T5) tornadoes that have caused damage associated with vortex winds between 62 to 72 m/sec. In addition, one 'moderately devastating' (T4) tornado occurred with winds between 73 to 83 m/sec. It is striking that the majority of these have been within the winter period during December (two events), January (one event), February (two events) and March (one event). The single month of the summer period that has experienced one of these intense events is September. So, the winter season may have fewer tornadoes, but it has produced the most threatening events in the recent Irish record.

A significant case in point was the T4 (62 to 72 m/sec) Carrigallen tornado that occurred in a quiet rural part of County Leitrim on 26 January 2002.[6] It occurred during the dark early hours of the morning, about 05.00 UTC, so there were no direct eyewitness accounts. But the winter thunderstorms that developed in many parts of Ireland that morning resulted from a deep moist air mass moving across County Clare from the south-west (Figure 7.4). They became quite violent as the airflow became strongly sheared in passing over County Leitrim's drumlin terrain, significantly aiding vortex development. A strong tornado left a trail of major damage across the fields and farmland. Whole mature trees were removed from the ground and displaced, pushed over and trunks snapped off; large gate pillars were snapped and tossed away. A track of intense damage up to 132 m wide was created (see Chapter 8). This was a single tornado. Despite careful enquiries over a wide area no evidence of other tornadoes occurring on this occasion could be found. The conditions that gave rise to this event were by no means unusual for winter months in Ireland.

Tornado intensity

In addition to the assessment of seasonal and monthly tornado frequencies that normally predominate when a national tornado climatology is developed, a comprehensive review of tornado intensity has become possible for Ireland. This is unusual in most countries because it requires a detailed on-site investigation for each event. But it is desirable because an intensity assessment adds a significant dimension to the climatic profile. This makes possible a more realistic assessment of the threat potential for an area when the threat is based upon measured impact details of past events rather than a theoretical analysis. In many countries intensity assessments are only available for a few tornadoes each year.

Over the twenty-year period up to 2017, all months had at least one tornado at the T3 intensity level, with the sole exception of July, which had a maximum intensity of T2. The low maximum intensity for July clearly demonstrates that tornado frequency and tornado intensity distributions do not go hand in hand. July is one of the two months that have the most tornado events overall. August is the other. But both July and August have had among the lowest intensity ratings for their tornadoes. For July, the maximum intensity for a tornado was T2. For August the maximum intensity was a T4 event. But more telling of their tornado profile is that the most frequent intensity rating for all of their tornadoes throughout the month was T0 for July and T1 for August. This is far from what many people expect of the warmest months of the year.

The pattern of monthly high tornado frequency and high intensity events not matching each other is further reinforced by the highest T values having occurred in the month of December. The most severe were events of T5 intensity, of which there were two, one in 1999 (24 December at Coolrain, County Laois) and the other in 2006 (31 December at Tolans Point, County Antrim).[7] But these intensities were equalled in years predating the twenty-year focus for much of this analysis. These were at Summerhill, County Meath (17 March 1995), Youghal, County Cork (13 February 1995), Foilmore, County Kerry (11 February 1990) and Belfast (26 September 1982). However, it has already been noted that the tornado event that appears to top all of these was on 24 January 1852, at Nenagh, County Tipperary.[8] It is striking that most of these occurred in winter months.

Overall, for the whole of Ireland, the most common intensity for a tornado recorded during this twenty-year period was a T2. It was also the most common event intensity in five separate months of the year (January, April, September, October and December).

Tornado clusters

Despite the preponderance of single tornado events throughout the year, it is a striking feature that a small, but significant, number of tornadoes occur in small clusters. This is contrary to the widely perceived pattern. This tends to be largely focused upon individual events on individual days. The manner by which most tornadoes are reported and processed by the mainstream media may help to explain this. A report received by a media outlet needs to be published quickly while the event is news. In practice, this quickly becomes the focus of attention and a wider picture is easily missed. So, if there is a cluster of events, it tends to be overlooked.

Such clusters consist of tornadoes, often accompanied by a number of funnel clouds, all occurring on the same day. An event of this nature has occurred on at least one occasion during each of the twenty years of detailed records. Daily clusters have occurred forty-eight times during these twenty years (Table 7.2). Only two months, February and March, have no record of such a cluster during the period, any tornado in those months having been isolated ones. Although the clusters have been spread over the whole year, they have been most frequent in July and August.

One type of single-day cluster occurred when a storm system maintained a suitable generative environment while travelling across Ireland. An extreme case was the tornado cluster of 25 November 2000, when site investigations established that four tornadoes occurred as the day progressed, from about 04.00 UTC at Brow Head in the extreme south-west of County Cork, to Donaghadee in County Down during the afternoon. In between, tornadoes developed as the storm system passed Rossmore in County Cork and Kilbeggan in County Westmeath, at different times during the morning. It is quite possible that one or more other vortices may well have been spawned along the way, but visibility was poor due to thunder, intense rain and hail on the one hand and, on the other, the fact that the system passed over land that in many places was peat bog, which was relatively empty of potential observers.

Sometimes clusters have developed in relatively close geographical proximity. Such was the remarkable case of a four-tornado cluster that occurred in a single county, County Westmeath, on 30 September 2001.[9] Each of these tornadoes initiated separately over a period of thirty to forty-five minutes. In terms of distance, the furthest apart were separated by no more than a mere 19 km.

The most numerous cluster for Ireland known to date was the six-tornado event on 1 January 2005.[10] On this occasion there were three separate tornadoes in County Westmeath, and one each in County Offaly, County Meath and County Armagh over a two-hour period. This will seem relatively slight by some international comparisons, but it does indicate that when building a database of this nature, the receipt of one report should prompt the question, 'where are the others?' In 2006 there were five such clusters. This was the largest number during any one of the twenty years up to 2017. Even though the other years had fewer clusters, they show that they are an important feature of the tornado profile for the country.

Monthly tornado days and funnel cloud days

Of course, the number of tornado days that have occurred each month is less than the number of tornadoes. But the difference is not great. Just as the monthly totals show August to have the most tornadoes, August is also the month when tornado days are the most frequent. Table 7.2 shows that 18 per cent of the twenty-year total up to 2017 occurred in that month. The earlier summer months of May, June and July, with 12, 12 and 14 per cent, respectively, also had more tornado days than other months of the year. However, the actual totals for each were relatively low, since over the entire twenty-year period August had twenty-five, July had nineteen, while June and May each had seventeen tornado days.

Most months have experienced a decline in the number of tornado days since the year 2009. The greatest decline has been for the summer months of July and August. This has been despite the intensification and expansion of the reporting community and increase in social communication, particularly via the internet. But proportionately other months have had similar declines, although the numbers involved are very small and cannot, therefore, be regarded as particularly significant, apart from describing the overall pattern.

Days when funnel clouds occur are usually days when there is the potential for a fully developed tornado. In many cases this does not happen. Normally, that outcome is only ever known retrospectively. However, this does not make the funnel cloud any less important. Its development is firstly a warning of what might be but, secondly, it is a potential danger to small aircraft flying at low levels. These include aircraft that are approaching or leaving airports, gliding and any other related airborne activities.

Funnel cloud days peak in the month of July (Table 7.2). This peak is a strong one since over a quarter of all funnel cloud days have occurred during that month. The summer months of June, July and August dominate the annual pattern overall, accounting for 66 per cent of the annual total. But equally distinct is the relative absence of funnel cloud days during the winter. The five months of November to March have recorded a mere eight days of funnel clouds in twenty years, just 3.5 per cent of the total. When these winter funnel cloud days have occurred, it has not been in association with tornadoes, not even on those days when there have been particularly significant tornadoes. For example, when the six tornadoes occurred on 1 January 2005, no funnel cloud was reported at all from any location within Ireland.

The time of day

Generally, weather systems move across Ireland quite independently of the time of day. However, their effects may be modified as a result of the daily rhythms of warming and cooling. While the sun delivers its highest energy load in the middle of the day and declines thereafter, the warming effects of this spread only slowly upwards from the surface. So, the highest temperatures in the air above the ground are felt during the afternoon hours. This increases any instability of the air, and it is likely to develop deep convective motions if other conditions are suitable.

There is a second part of this daily process that may also come into play. This is linked to the cooling processes that occur towards the end of the daylight hours as the sun declines and its energy input is lost. Then, the atmosphere begins to cool down as it loses some of the energy it has stored up. However, the way this happens has some surprising effects. One of these is that instead of dampening down the instability levels and convective motions initiated during the afternoon, it may actually increase them. On a late summer's afternoon, clouds may build up and then, after dark, a night-time thunderstorm occurs. This is a familiar sequence throughout Ireland. In many cases what happens is the growing clouds blanket in the warmth of the afternoon and, as the sun declines, cooling occurs, but from the upper part of the cloud structures. This increases the temperature differences between the surface and the middle of the atmosphere. A resulting strengthened vertical temperature gradient encourages even greater buoyancy and instability. Heavy showers and thunder may follow if other conditions are suitable.

Table 7.3 The most common hours for tornado initiation and the earliest and latest times per month (UTC), from 1980. Italicised bold times are events that initiated during the night.

Hourly Tornado Profile												
Diurnal Extremes	J	F	M	A	M	J	J	A	S	O	N	D
Earliest	0840	0720	***0245***	0700	1030	***0215***	***0210***	***0040***	0700	***0610***	***0400***	***0050***
Peak Hours (from … to)	1100						1100			1100	1100	
								1300		1300	1300	1300
	1400		1400	1400								
		1500			1500	1500						
		1600	1600	1600					1600			1600
							1700					
					1800							
									1900			
						2000		2000	2000			
Latest	***2000***	***2045***	1850	***2030***	2100	2030	***2230***	***2350***	1925	***2245***	1400	1700

The contribution of these processes is evident in the pattern of tornado occurrences in Ireland. This shows there is a distinct afternoon peak frequency. Table 7.3 shows this peak occurs anywhere between 11.00 and 18.00 hrs, although the individual hour with the maximum occurrences is that between 16.00 and 16.59 hrs, when just over 13 per cent of the annual total occurs. Nevertheless the other hours of that peak period were not far behind, since each hour between 11.00 and 20.00 hrs contributed a further 7 to 9 per cent each. As a result, some 74 per cent of all tornadoes occur in that eight-hour time span.

Night events appear to contribute very little to the overall total. This may seem to be expected if a simplistic view of the role of instability in tornado formation is taken. But the complex range of conditions that can produce a tornado mean that when instability is relatively low tornadoes may occur, even at night.[11] Table 7.3 shows that in seven of the twelve months the earliest tornado formation of the day was in the dark early hours of the morning, while six months had tornadoes initiating late in the day after sunset. This is certainly far more than is generally expected.

Reference has already been made to some of the significant nocturnal tornado events. The tornado at Carrigallen (mentioned above) is a case in point. But it should be noted that the record for such events is inevitably very limited because they cannot be spotted in the dark. They have only been recorded at night when they have occasionally caused damage or injury. The Irish record shows only thirteen night tornadoes over twenty years, although there are more in the record prior to 1998. There is no preference shown for summer or winter months, such as features in the UK data.[12] However, Ireland is generally sparsely populated, so the likelihood of any general awareness of these events during the night is very low. But as the growing population of Ireland spreads out this may well change. Even during the daytime, the investigation and identification of many tornado events has only been possible because they have been seen by rather than directly impacting an observer.

The spread of tornado events through the twenty-four hours of the day is very wide. It shows that there is little difference between winter and summer months. There is a small difference with regard to the slightly later times of peak activity in the months of May to September. However, there is no simple, clear distinction between winter and summer months. The overall pattern points to the probability that throughout the year tornadoes may occur at any time of day or night.

Climatic chronologies and forward projections

The chronology of extreme weather events in databases spanning a few decades is very helpful, but far from definitive, in determining the probable frequency of tornadoes. If the chronology was of a random series of events over a very long period of time (as assumed in most statistical tests), then the events would not be equally spaced, nor would they necessarily be of the same intensity range. There would be periods of clustering and periods of relatively low frequency. So, annual totals would not be the same. Their fluctuations could simply be an effect of randomness. Likewise, tornadoes of a higher intensity than appear in the more limited database would likely occur, compared with the range of intensities included in a shorter chronology.

So, when annual tornado frequencies or event intensities occur that are outside the recorded range, it need not be due to a fundamental change in the climatic regime. Indeed they are to be expected – at any time! The extremely important conclusion from this is that the potential threat to be managed is greater than that suggested by the limited record. Nevertheless, the database for tornado events is the first expression of the risk they pose. Irish data shows that the probability for different levels of occurrence in Ireland are as follows (Table 7.4):

Table 7.4 Probabilities for annual tornado and waterspout totals being equal to or greater than given frequencies.

Number of tornadoes and waterspouts/year	Probability of being equal to or greater than the number
2	1.0
3	0.95
4	0.95
5	0.95
6	0.80
7	0.75
8	0.55
9	0.50
10	0.45
11	0.40
12	0.40
13	0.25
14	0.20
15	0.15

As referenced earlier, for Ireland as a whole, the statistical average is 9.3 tornadoes and waterspouts per year which, for convenience, gives a rounded figure of 9. The data can also be used to show the probability of anywhere in Ireland being hit by a tornado, assuming all areas are equally vulnerable to the threat. Excluding coastal and marine events, the average damage path length is 4.2 km, and the average path width is 46.9 m. This detail can be used to estimate how long it would take for every square kilometre in Ireland to be hit by a tornado – approximately 3,234 years!

Generally speaking, although statistical probabilities such as these give a technical expression of the best estimate of what might be expected, they mean little to ordinary people. Public expectations are largely based on personal experience and additional information that makes sense. Indeed, often personal experience seems to contrast with careful statistical analysis, as events near Dungarvan, County Waterford, demonstrate.

The Dungarvan tornadoes

The weekend weather of the last week of September 1978 had been bad. At sea the daily Le Havre to Rosslare, Fishguard to Rosslare and Pembroke to Cork ferries had been seriously delayed by heavy winds and very high seas, which also brought the fishing fleet along the south coast to a complete standstill. But on the Sunday afternoon this was a mere distraction as most of Ireland settled to listen to the All-Ireland football final between Kerry and Dublin. Dick Hanrahan and his family had settled down to listen to the match in their Ballinacourty farmhouse near Dungarvan. Suddenly above the chat of the commentary team came quite a different sound. First it was distant, but within seconds it grew to an alarming crescendo as the roar of a sudden storm seemed to engulf the farmhouse. Outside the world seemed to have gone crazy as ferocious spiralling winds cut across the lane and missed the farmhouse by about 50 m. However, the outbuildings had not been so fortunate. A large two-span hay barn had been virtually demolished and the roof of a recently built modern milking parlour was ripped off. Parts of it were carried over 200 m, their walls damaged, and there was debris flying all over the place. Viewed from the house it was chaos. But the chaos soon resolved itself into a funnel of a tornado as it headed northwards parallel to the coast where it disappeared out of sight of the terrified onlookers. So a new barn had to be built.

But years later it happened again – the same barn was taken out by another tornado! As if in defiance of averages, statistical probabilities and calculated return periods it happened on 1 June 1998, just 22 years, eight months and four days later! This time it was the early hours of the morning when the tornado struck. Again, the farmhouse escaped, but the barn was structurally damaged. The severity of this event can be judged by the near fatal experiences in a nearby caravan park on a local farm. It would have been a spectacular sight during the daytime, but completely invisible at night, to see the waterspout develop just offshore in Dungarvan Bay. So, there was no warning. But two families knew that they were in deadly peril when, just before 02.00 UTC, they were suddenly pounded by winds that shook their mobile homes to their very foundations. One family had their world literally turned upside down. They were lifted far enough off the ground to turn over with ease as the mobile home was carried from its placement to the edge of the site, where it hit the ground. The family was flung all over the mobile home as it twisted in the air and was dumped on the ground, shattering it. Fortunately no one was killed, although some of the nine holidaymakers had severe injuries. The tornado moved onwards across the fields, depositing debris and scouring crops, before reaching the Hanrahan's farm.

The likelihood of a tornado striking the same place twice is extremely remote. Yet it happens. But it is not at intervals of 3,324 years as the statistic above might suggest. Statistical definitions of climatic threats such as a tornado are very inadequate instruments for either defining expectations or describing the past in terms of weather events. Indeed, statistics may give a false sense of security for infrequent, but intense, weather events.

The atmosphere as a chaotic system

The complexity of interactive processes in the atmosphere results in their outcomes being largely unpredictable on most time-scales. Perfectly accurate predictability requires atmospheric and other related environmental information, as well as a comprehensive understanding of their interactions, in two areas. Firstly, knowledge of the initial state of the atmosphere in perfect detail is needed. Secondly, a precise understanding is needed of exactly how those conditions are modified step by step in order to produce an outcome as a distinct meteorological phenomenon or a climatic pattern (when multiple such events occur over a given period of time).

For this reason the atmosphere has been described as a chaotic system. This is not the chaos of popular thought, which uses the term to refer to a state of disorganised, unpredictable confusion. The chaotic nature and behaviour of the atmosphere refers to its multiple interactive complexities, behind which there are patterns and fully deterministic outcomes. But it is very often extremely difficult to recognise those outcomes. Any imprecision in the description of the initial state, however small, grows very quickly with time and the number of possible phenomena resulting from this increases accordingly.

This was stumbled upon by Edward Lorenz in the 1960s when he found that weather prediction is highly sensitive to initial conditions.[13] Minute inaccuracies, or any tendency to use an approximate value (such as rounding a value because of multiple decimal points in the true value), would quickly have an effect that was so magnified as to make the outcome totally different from what it would otherwise be.

The so-called butterfly effect is familiar to many. This was a description used by Lorenz to illustrate how a small change in just a single condition of an atmospheric system can result in very large differences in a later state. The example used was a butterfly flapping its wings in China potentially impacting the development of a hurricane in Texas. It caught the imagination of scientists worldwide and vividly captured the importance of chaotic processes.

Thus, explaining meteorological events or climatic patterns they produce, or even forecasting them, with any certainty, is not possible.[14] The levels of observational detail and accuracy necessary for this is simply not possible for the foreseeable future. There is a long way to go before this can significantly change. But building databases of weather events that we want to understand and to have a protective response ready for those events that pose a hazard, are small steps towards that end.

There is a sufficient run of data to show that simple year-on-year predictability does not work, whether based on linear or non-linear trends. Nor can any trend line be constructed through the data that would have any reliability for the frequency of future tornado events. However, it seems to be reasonable to expect that the majority of the time periods for which data is presented will, in the immediate future, experience similar frequencies. But this is not a calculated prediction for the future, merely a projection of past event frequencies.

CHAPTER EIGHT

Investigating Irish Tornadoes

Overwhelmed by Nature's fury
Split asunder
Flung afar
Driven deep
Patterns amidst the chaos.

Standing in the middle of a devastated countryside with once proud trees that had stretched 20 to 30 m into the air now lying shattered on the ground may be part of a site investigation, but it is also often an awesome experience. These vivid images of destruction leave an imprint on the mind of the atmosphere's irresistible power and ability to leave its mark and to reshape the landscape over which it surges. In rural Ireland, these marks last a long time. They yield a great deal of valuable information about the tornado that left such an impressive footprint.

In Ireland, most tornado research has to be based on the investigation of the tornado track after the event has occurred. Field observation and site investigation are central to this task. There are three key ways in which site investigations have become important:

1. **To verify the event.** It must be established with certainty that it happened, and that it was a tornado rather than several other possible wind phenomena. This requires establishing the detailed characteristics of the event. There is no other way to do this. There is no system of storm watchers and chasers that follow developing storms, reporting and recording their development as they do so,

including the tracks they follow, the changes in direction, shape and intensity that they undergo. Neither is the radar technology that monitors and records the detailed evolution of storm cells generally available for this work in Ireland, although the availability of finer resolution radar has improved in recent times. These different sources of data yield their own particular information for each event.

2. **To develop theory.** Field observations have always formed the basis for understanding tornadoes by developing theories and models to understand what produces such violent atmospheric conditions. Even in countries where radar and other remote sensing technologies are well developed for tornado studies, ground surveys, based on enumerating and mapping damage and other details of a tornado's path, are still an integral part of tornado investigation.

3. **To study the range of impacts of a tornado.** These vary from the destruction of elements of the physical landscape and built structures to the characteristics of the debris field. But they also extend to the trauma and psychological disturbance caused by the experiences of people and their local communities. Traditionally, tornado studies have largely ignored many of these.

The Tornado Footprint

The tornado's footprint on the landscape comes in many shapes and sizes. It poses many challenges for the site investigator, both in data collection and interpretation. Four case studies of recent tornadoes in Ireland show how diverse these challenges are. Each of these tornado events were highly significant in different ways.

The Banemore tornado[1]

Walking across the Stack's Mountains on a clear summer's day the visibility is stunning. Not only the distant visibility, but the detail of the plant communities that make up the boglands and rough grazings that cover the mountainside as far as the eye can see. The views are mostly unbroken by any forestry or clumps of trees. Here the wind could have its fullest dominion, flowing over it all without any hindrance. On that summer's day in July 2000, when the investigation of the event at Banemore commenced, there was not a hint that one of the largest tornadoes recorded in recent years in Ireland had swept across the area only four days previously. Then, every plant had been flattened by the blast and every creature had sought refuge in their burrows beneath the surface. For nearly ten minutes the tornado had moved very slowly across the terrain. Then, the swirling winds within its vortex near the ground may have reached well above 160 km/h.[2] All that destructive energy within a diameter of a little more than 150 m! Now the landscape was very calm, as if nothing dramatic had ever happened. But there had been eyewitnesses, photographs and videos of the event, even though now all seemed to be in its place. To the human eye nothing had been damaged, or even disturbed. Later that day even a local countryman, searching for traces on the land he knew so well, drew a blank.

This was not a unique experience. One of the most frustrating aspects of investigating tornadoes in Ireland is the nature of the Irish environment itself. In some regions much of it is ill suited for recording the passage of a tornado. Many environments are well used to and well adapted to extreme winds. Plants grow low on the ground and the surfaces of bogs are compact and plastic-like, not vulnerable to scouring by the wind or wind-borne objects. In those places where the tree comes into its own and the vegetation becomes both taller and more brittle, the extreme wind speeds of the tornado may snap trees like matchsticks and batter them down. Even so, many have developed wind-denying strategies, such as in their rooting systems and their above-ground flexibility. The most vulnerable to severe wind damage are artificial structures. But the few in those regions are either already so ruined that it becomes very difficult to recognise new damage, or very strongly built to withstand the winds of mountain and coast, hilltop and exposed plain.

The Galway–Roscommon tornadoes

In the darkening evening of 2 October 2013, and into the next day, reports began to surface of tornadic activity in various locations across counties Galway and Roscommon. Serious damage had occurred to homes, farm buildings and large trees. The reports came from an area between Meelick in east County Galway to Kilroosky in north County Roscommon, a distance of 42 km. This had occurred between 17.30 and 18.30 UTC

8.1 The wedge tornado crossing counties Galway and Roscommon. (Photo by Laurence Cheeseman)

(18.30–19.30 hrs), well into the hours of darkness at that time of year. Subsequent site investigations and eyewitness evidence pieced together a number of different damage tracks and accounts of tornado activity.

It all began at 17.30 UTC (18.30 hrs) when a tornado was observed close to Meelick weir on the River Shannon. It intensified as it crossed farmland into the Clonfert area, where it did major damage, before continuing northwards across extensively cut boglands. There it left little trace except for depositing some of the debris from damaged trees and roofing materials. However, along this route it had left a trail of significant tree damage. More seriously it had destroyed a lived-in mobile home while a young family was in residence, damaged other built structures, especially roofs, and memorably demolished headstones in the graveyard of the historic Clonfert cathedral. From a comfortable distance of several miles this was well recorded on film and video in the diminishing half-light of the winter's evening (Figure 8.1). But the video evidence provided no ground details and would have been somewhat uncertain as evidence if site information had not been available. As it was, the traceable track of 10 to 11 km petered out going into the bogland and it cannot be certain how much further the track extended. However, eyewitness and photographic evidence was at least sufficient to show that somewhere over the bogs the funnel did dissipate.

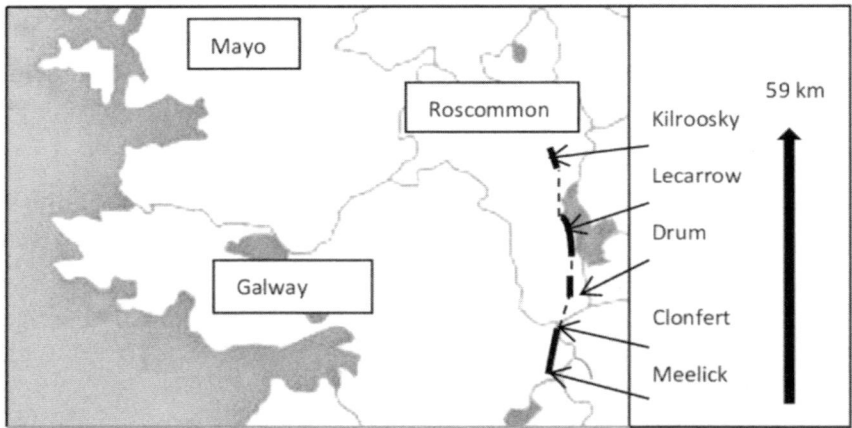

8.2 The tracks of the County Galway–County Roscommon tornadoes.

Then, around 18.00 UTC (19.00 hrs), about 20 km north of Clonfert, another fully developed tornado emerged out of the bogs close to Drum and quickly swept through the village northwards to Monksland, which is part of the western suburbs of Athlone. At the time the rain was torrential, and it was almost dark. But the funnel was still observed trekking across the fields towards Monksland, where it did significant damage to the rooftops, sweeping tiles, guttering and other materials away into the air. That track was only approximately 5 km in length before it also petered out (Figure 8.2).

The next tornado arrived in total darkness. As a waterspout it had tracked along Lough Ree for an unknown distance before coming ashore at Carrowmore and moved northwards beyond Lecarrow in northern County Roscommon. At Lecarrow it severely damaged a number of large trees and cut a spectacularly clear swathe through a maize crop. As one landowner and his family had sat down to their meal at 18.30 UTC (19.30 hrs), intense rain drove against the house before there was a deafening sound, 'like a jet plane' they said, blocking out all other noise. It was not until the following morning that the severity of the damage became apparent, and the track was discovered. During the site investigation more damage was found, all of which pointed to an event with an intensity of T3. Further north it imposed its final known stretch of impacts near Kilroosky, where damage was minimal and the impact slight.

Had it been assumed that these four tornado tracks were a single continuous one that crossed the featureless boglands between the areas of intense damage, the total track would have had a length of 59 km. This would have been the longest on record in Ireland. But the investigations established that separate tornadoes formed either on water or came from the boglands, into which at least one of them disappeared. A single storm cell appears to have been involved in generating this situation. The tornado was preceded by intense rain and was positioned towards the rear of the cell close to the intense rain.

Distinguishing between a single long track and several shorter tracks from a number of discrete tornadoes has been a challenge in a number of cases. In particular, the St Patrick's Day tornado of 17 March 1995 is recorded as having a damage track of 24 km, through County Westmeath, County Meath and County Dublin.[3] But uncertainties about its continuity and the involvement of a single tornado remain. It is quite possible that there may have been more than one.

The County Westmeath tornadoes[4]

This all began with an alert issued by TORRO on Friday, 28 September 2001. However, there was nothing definite in what was an experimental forecast. In the final analysis this was an opinion based on a couple of broad indicators and the considerable experience of the forecaster, Tony Gilbert. Being the first experimental forecast specifically for Ireland, it is worth noting its relevant parts. Drawing attention to broad official model forecasts on the web pointing to south-west Ireland getting the worst of the possible severe weather expected on the forthcoming Sunday, he stated that, 'it will be in this region that unstable air and intense 500 mb vorticity might be present' and went on to say,

> If the forecast is right then a surface secondary triple point low may form within the circulation of the main low pressure of 960 mb over Southern Ireland. The tightly coiled wind vectors shown at the surface combined with a possible 160 kt upper winds will likely increase wind veer and shear dramatically and could be a set up for tornadoes!

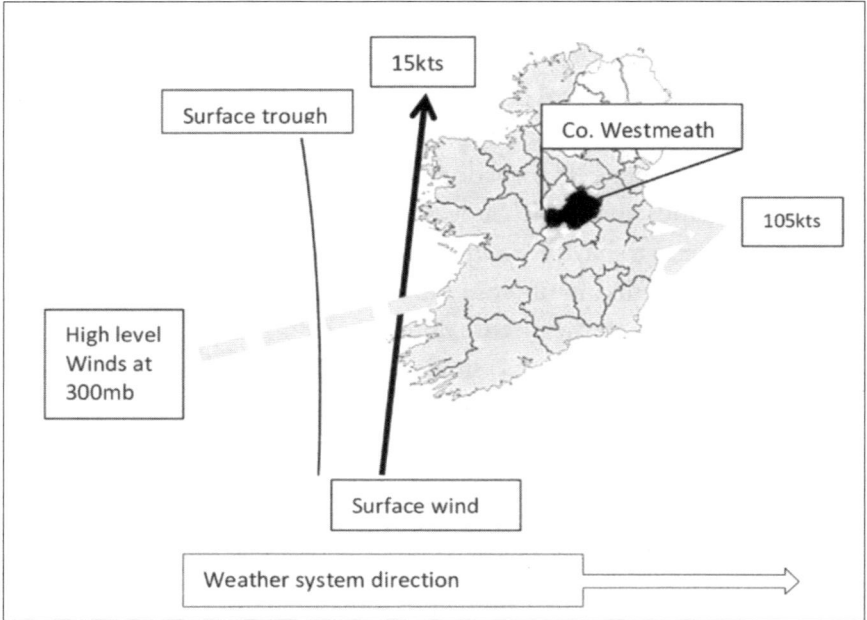

8.3 Significant features from the synoptic charts for the Westmeath tornadoes, 2001.

Keeping an eye on forecast charts available on the web, the weather appeared to unfold during the weekend much as the TORRO forecaster had expected. But Sunday became a disappointment. The County Cork skies showed growing cumulus congestus clouds passing overhead, but anything other than a heavy shower of rain seemed unlikely. Nor did news bulletins carry information about severe weather elsewhere in the country.

From the synoptic charts some of the significant ingredients for a tornado event were apparent. These have been extracted from the general surface synoptic chart and the Valentia upper air sounding for 12 noon in Figure 8.3. They showed a strong vertical wind shear for both wind speed and wind direction, a shallow layer of instability between them and a surface level trough in atmospheric pressure approaching Ireland from the west, where the surface wind could converge. All these factors coming together as the system moved westwards across Ireland would create a promising environment for a tornado. However, the highly generalised nature of these data and the scale of analysis is such that what seems to be promising is often disappointing. So it seemed yet again.

But the next day the persistent curiosity of several people changed all that. Among them was Hans Temperley, of Ardmurcher House, just outside the small village of Horseleap, in County Westmeath. Severe damage had been done to large mature trees in a small area of his property, about which he was very curious. A second was Dick Hogan, the editor of the *Westmeath Topic*. He was reporting on a local storm that did significant damage in the town of Mullingar, also in County Westmeath. But he did not consider it to be an 'ordinary' storm and took his enquires further. Both men were quite unaware of what others in the area had experienced. From their brief accounts it became apparent that it was important to investigate what had occurred that Sunday evening, because a tornado was a distinct possibility.

The site investigation began at Ardmurcher House, where the view of trees ripped apart was striking. From driving around nearby lanes and making subsequent visits to places where particularly intensive damage was spoken of by local people, it appeared that a tornado had forged its way across the countryside to create a damage track of about 9 km, but this was some distance from Mullingar. However, from further investigations there it was apparent that the damage at Mullingar had also been caused by a tornado. Between the two damage tracks there was a gap of about 15 km. Although this gap was almost in a straight line, there was little identifiable storm damage along it, but the conclusion that the Mullingar tornado was a separate one was confirmed by the timing of the events, as reported by local people.

However, the distribution of damage in the Horseleap/Streamstown area was not simple. Sometimes it is the small details that do not quite fit that provide a clue to the real situation. So it was in this case. Accommodating an outlier of damage west of Streamstown required the tornado to have taken a very strange twist. This appeared highly improbable. The urgency of completing as much of the site investigation as possible in a single day had led to the adoption of a damage cluster method of analysis, whereby a number of what are regarded as key local spots of intense damage are investigated and inferences are then made about the areas in between. As a result of this, an approximation of a single tornado track was made, but a cloud hung over this conclusion. There were nagging doubts because some of the damage details did

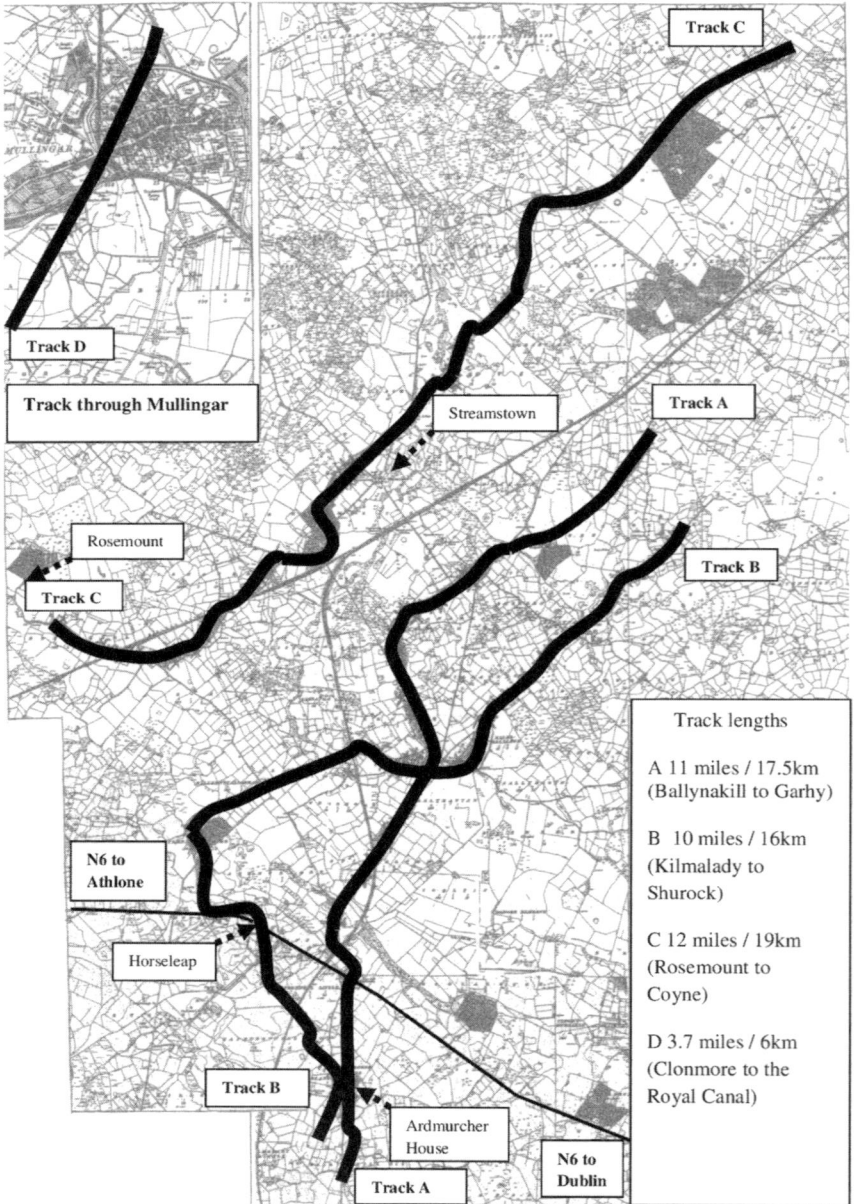

8.4 The tracks of the County Westmeath tornadoes.

not quite fit. As a result, a further, more detailed, site investigation was carried out some days later. In the meantime new eyewitness information became available and the involvement of some senior pupils of Kilbeggan Secondary School, with the assistance of their deputy principal, significantly added to the field evidence.[5]

The results of the new investigation were a revelation. It was clear that in the Horseleap/Streamstown area there had been three separate tornado funnels, clearly identifiable by almost continuous wind damage to trees and a small number of built structures. The reconstructed tornado tracks mapped in Figure 8.4 show that two of these had their beginning close to Horseleap and one near the village of Rosemount. This was the first time that such a multiple event had been recorded and confirmed in Ireland. Indeed, if the tornado in Mullingar was included, and it was clearly a part of the same storm system, there was a total of four separate funnels that had tracked across the Irish countryside that Sunday evening.

This would have been an amazing sight. But the time between 17.00 and 18.00 UTC (18.00 and 19.00 hrs) on a Sunday evening was not a good one for outdoor activities, because one of the most popular TV programmes, *The Simpsons*, was scheduled for that time! Otherwise, if not sheltering from the weather, other people were indoors for Sunday tea. Most of these people only realised the storm was severe when their power failed (due to power lines being brought down) and they heard the maelstrom outside their homes. So there were few eyewitnesses. The few that were fortunate enough to see something gained no impression of the wider picture of events. Traffic on the Dublin to Athlone road (which was crossed by two of the tornadoes) came to a standstill at a distance because of the torrential rain and winds that shrouded the entire area. The drivers only saw the debris afterwards as local people struggled home through lanes blocked by fallen trees. So most of the events were invisible to the local population and the fuller picture was revealed only when the damage details were put together and plotted, tree-by-tree and field-by-field.

Further analysis of the track patterns and other data clarified the nature of the funnels that had caused the damage. Two funnels had started their tracks to the south of Horseleap and had then separated. They may well have started as a single funnel but became two separate entities in the grounds of Ardmurcher House. From that point two separate tracks emerged and crossed the fields, one to the north that went directly towards the main road which it crossed close to the Horseleap Garden Centre, and the second to the north-west where it took a longer route across the fields. This behaviour has been recorded elsewhere as the result of suction vortices of the main tornado.[6]

If the suction vortex theory is correct, a single tornado close to Ardmurcher House would have had a strong central downdraft that penetrated to the ground surface. This would separate off another vortex that would then rotate around the primary vortex. Since the damage pattern shows that the same two vortices approached each other very closely after about 4 km, this seems to be a reasonable interpretation: they do appear to relate to each other. But it is impossible to identify one of these as the primary vortex of the two, and suction vortices are normally an order of magnitude smaller than the parent vortex. It is not even certain that both tornadoes occurred at exactly the same time. There is also a small amount of evidence that they may have had a separate existence a small distance to the south of Ardmurcher House. Therefore, the vortex theory cannot be confirmed, and it has to be assumed that two separate tornadoes occurred near Horseleap.

However, tornado C in Figure 8.4, the Rosemount tornado, does appear to have been independent. It started 8 km north-west of the two tornadoes that came from the Ardmurcher House area, although it did approach the track of the Horseleap tornado by 4 km.

This whole sequence was one of the most complex Irish tornado events that has been investigated. There have been other occasions when a number of individual tornadoes and funnel clouds have occurred over a very wide area of territory. But on this occasion these multiple vortices were spawned within a period of about thirty minutes at most, over a small portion of County Westmeath. The investigation highlighted the importance of detailed surface investigation and the danger of assuming that tornadoes come singly.

Tornadoes differ markedly in their size, shape, life history and lifespan. These factors leave their own particular marks on the ground as features of the tornado's path. They express themselves in the path length and width as well as in the specific details of the damage. Site investigations are, therefore, carried out to accurately determine these features.

The Dublin Bay tornado [7]

A tornado in, or near, Dublin is news on a national scale. Many that have occurred across the country have not been considered newsworthy, at least by the national media, and have not received coverage. But Friday, 18 August 2001 was different. The news broke on the radio on Friday evening. People travelling home from work through Dollymount reported a tornado over Dollymount Strand, others that a tornado had hit the Bull Wall that separated the Strand from the approach to Dublin Harbour, and others that it had turned over a Land Cruiser while it was being driven on the Strand. Immediately the media responded. Was this a near miss for the city of Dublin itself?

The development of atmospheric conditions to produce these effects is indicated in the synoptic charts for 12 noon that day. Figure 8.5 shows extremely strong vertical wind shear at a low altitude in the Valentia sounding. Wind speed was a mere 5 knots at the surface and 20 knots above it at 600 hPa, but the directional shear was a formidable 90 degrees. From a surface direction of east-south-east it moved to south-south-west at 600 hPa. CAPE was very small at 16 and was concentrated in a small layer where the wind shear occurred. Such a low level of CAPE would not ring any alarm bells, but a surface trough of convergent winds approached from the south, which could have enabled a limited amount of vertical stretching. However, even when combined, there was little that was definitive about these conditions for vortices in Dublin Bay.

Initially the multitude of witnesses seemed to prove that a tornado had occurred. But the ultimate proof for many was the TV coverage the next day. This had an extract from a video recording made by a couple from the USA on a brief visit to Ireland, who happened to be at hand when the event occurred. The TV item consisted of three shots. The first showed the cloudscape over Dublin Bay clearly developing a funnel cloud as it descended some distance beneath its parent cloud. The next went to an upturned Land Cruiser being rescued from Dollymount Strand, with overriding commentary that this was damage from the tornado. Third, was a comment by a representative of Met Éireann that in the right circumstances tornadoes can occur in Ireland. Put all that together and the case seemed to be pretty convincing to most people.

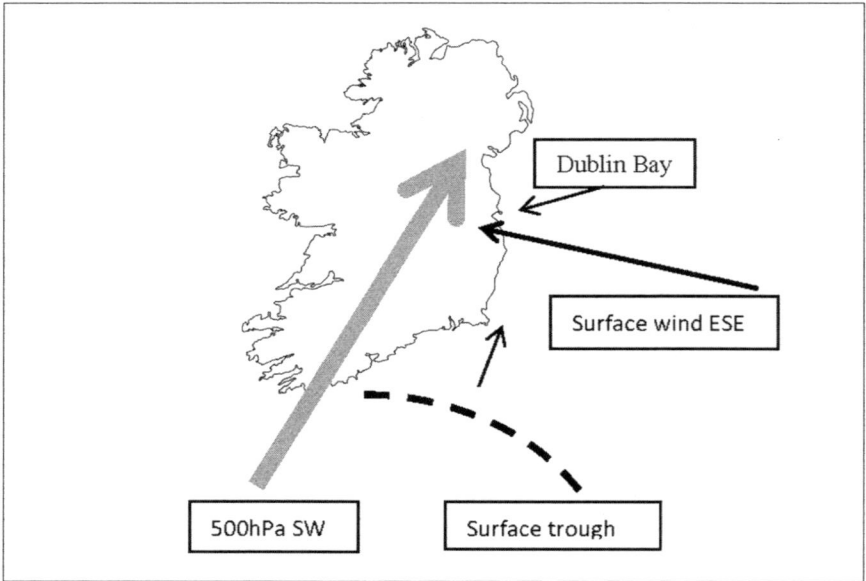

8.5 Significant ingredients from the synoptic charts for 17 August 2001.

However, since Mr and Mrs Nolan from California had returned to the USA on the day following the event, an opportunity was sought and kindly granted from their Californian home to analyse the full video. This showed something different. The sequence of events began with views of Dublin Bay and the upturned Land Cruiser. It does not show how the vehicle became upturned. The operation, involving Gardaí and a rescue truck, was the centre of attention for people on the beach for some time, as one would expect. After about 40 minutes (according to the timer shown on the video), as the couple were considering leaving the strand, they spotted rotation in a cloud out over the bay. This was seen to develop into a funnel cloud which descended some distance below the parent cloud.

The funnel remained a considerable distance out over the sea throughout the period of the recording. There was no evidence of disturbance on the sea surface that suggested the development of a waterspout in association with the funnel. Even if there had been a waterspout, and even if it had proceeded to the shore and across the Strand, by that time the Land Cruiser had already been removed. What had been presented to the nation was a piece of rapid editing based upon a wrong assumption. As a result, when the Dublin tornado is spoken of today, very often it raises the response, 'Didn't that turn over a vehicle on the beach?'

What really happened was a sequence of events that became mixed up in reporting. They were only sorted out after the sifting of many eyewitness accounts. As large cumulonimbus clouds developed during that afternoon they produced, at different times, a total of three waterspouts and a funnel cloud. None of these moved onto or crossed over Dollymount Strand. Instead, there was a surface-based dust devil/land devil that grew to a considerable height, about 20 m, on Dollymount Strand. This travelled from the Bull Wall area towards the Royal Dublin Golf Club and beyond.

Stages in the Development of Tornado Tracks

Although the tornado's footprint on the landscape does come in many shapes and sizes, there are a few features that characterise most events. As tornadoes develop and mature through their life cycle, so their imprint in terms of damage changes. Sometimes this is sudden. Much of the detail depends upon the horizontal speed at which a tornado travels across the countryside, the intensity of the vortex and the nature of the surface over which it is travelling.

A generalised life cycle hypothesis for a tornado was originally outlined in the USA.[8] This was based on its structure, appearance and the dimensions of the track it produced. The four stages of the life cycle each had a different expression in the swathe of damage along the tornado's track. In Ireland, actual studies of tornadoes have identified that at least three stages can be identified in tornado development, where tracks are often simpler. The Carrigallen tornado illustrates these three stages well.

The Carrigallen tornado[9]

On 26 January 2002 a tornado occurred at Carrigallen in County Leitrim during the early hours of the morning. As mentioned in Chapter 6, there were no eyewitnesses because it was before dawn, although some local people were very close to the storm and heard it quite clearly. Trees were the main target available to the tornado, although a few structures were damaged as well. These made up a track with a distinctive pattern, which, although simpler than the USA model, is more typical of tornado tracks in Ireland.

The field-by-field site investigation of the damage track coded and mapped the severity of damage to each tree and damaged structure. The result showed a track of 2.4 km and a maximum width of 132 m. An average track width of 51 m was derived by measuring the damage swathe at each field boundary. Based on these results the three stages in the development of the tornado are shown in Figure 8.6 and Table 8.1.

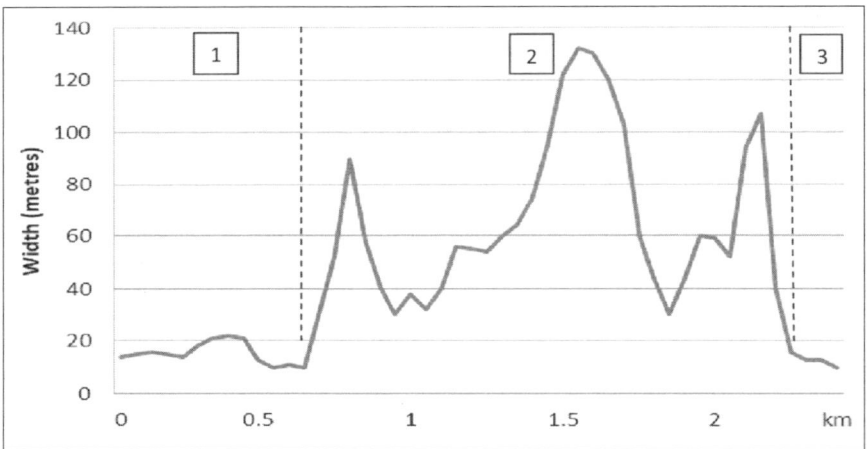

8.6 The width of the Carrigallen tornado track and the three development stages.

Table 8.1 The Carrigallen tornado: Damage details in each stage of its development.

	Track length (metres)	Mean width and range (metres)	No. trees damaged (type 1)	No. trees damaged (type 2)	No. structures damaged
Stage 1	660	19 (15–26)	9	22	0
Stage 2	1518	69 (37–132)	56	37	3
Stage 3	222	12 (10–16)	2	2	1
Totals	**2400**	**51 (10–132)**	**67**	**61**	**4**

type 1: Whole trees blown over or trunk snapped
type 2: Damage to tree branches only
type 3: Structures damaged

Not only are there clear differences between the three stages shown in Table 8.1 but the transitions between them are quite strongly marked on the ground. The first stage of the tornado had a track up to 26 m wide, although it was mostly less than that. Then, over a space of only 63 m, the track width grew to 69 m as it entered stage 2. This also marks the start of the first area of severest damage. From this sudden widening and for the next 200 m, a mere two fields, there were eleven trees with type 1 damage and twenty-four trees with severe type 2 damage. But Figure 8.6 shows the track's width fluctuated strongly. Further along the track is the transition between stage 2 and stage 3. This was equally rapid. Over 116 m the track width declined suddenly from 111 m to a mere 16 m. So, the three stages may be summarised as follows:

Stage 1. Initial growth. This is the start of the track, where the damaged area is often at its narrowest. However, the damage is not always at the surface. Where vegetation or structures are relatively high, very frequently the start is marked by damage above the ground only, as the descending zone of high-speed rotation reaches critical limits for damage to the trees and high rooftops. When the surface is more open the whirlwind rotation will be at the surface and will possibly produce a low-level whirlwind of loose materials such as dust or leaves, though this does not always happen. It is dependent on the underlying surface and the weather conditions preceding the event, for it is harder to raise dust on a muddy field than a dry one and on wet rather than dry leaves. Overhead, a funnel cloud is evident. But it is not unusual for the tornado funnel to appear incomplete as condensation will not have occurred throughout the length of the rotating vortex.

Stage 2. A middle mature stage. This is highly variable in width. It also varies considerably in terms of its contribution to the total track length. It can form the majority of the track length, but not always. Its variability is partly related to intensity fluctuations as the tornado journeys. In some cases, there are marked changes in the direction of the track. Debris accumulation is a feature. This largely accumulates from the direct impact of the vortex winds. But often the inflow winds are strong and also contribute significant debris as well, further widening the track.

So during this stage the damage area tends to widen. In addition, it is not uncommon for debris on the ground to show different directional alignments. If there was no fully developed condensation funnel from the parent cloud to the ground in stage 1, it has developed by this stage. Here the tornado reaches its largest diameter and the debris cloud associated with it grows much higher as well as wider. Conditions during this mature stage are rarely constant. As a result, the characteristics of the footprint can vary greatly.

Stage 3. Decline. This is a shrinking, fragmenting stage. The damage track may become much more conspicuously fragmented. This can be marked when the funnel has become stretched by the parent cell and has become less aligned to the surface position of the funnel. This appears to be a common effect when the cell moves too rapidly for the funnel to keep up. Sometimes the funnel ropes out, but it is not unusual for it to still be destructive as it does so. When the vortex is quite narrow, the vertical winds may become stronger relative to the horizontal winds. Small plants, for example, may get pulled out of the ground rather than flattened, in such an environment. As the condensation funnel thins, so the path narrows. As the funnel retracts or breaks up, the first to benefit is the ground surface. In reality, the vortex structure may remain for some time, but the rotation speeds lessen and the atmospheric pressure at the surface quickly recovers.

In each of these stages there tends to be a vertical stratification in the damage pattern. At the surface are the lowest wind speeds. But they are often able to carry loose debris (large and small) horizontally towards the core of the vortex. This is subsequently lifted within the vortex as vertical winds predominate over horizontal winds. At the surface, all structures, both natural and engineered, are strong and at their most secure. They are at their most vulnerable when there is a structural discontinuity. But above the surface zone, wind speeds increase, and the security of structures declines with distance from their fixed base. Horizontal winds will reach their maximum above the level of most surface structures vulnerable to damage. A further complicating factor to these simplified damage track patterns occurs where the underlying surfaces influence the development and track of a tornado, as in mountainous terrain, or where there are sudden changes in surface roughness and within urban surfaces.[10]

One of the noticeable deviations from simplified tornado damage track patterns is where the tornado track appears to be fragmented. For some this is evidence of a tornado skipping. Such an erratic or discontinuous linear damage path can mean a number of things. Either there was continuous contact between vortex and ground in the path, but it was too weak to do damage; there was nothing to damage; similar structures were better constructed or the orientation of the structures resulted in varying degrees of vulnerability.

Where long breaks in a tornado's path have occurred, even when those breaks are aligned to other damage tracks, it has often been found that more than one tornado was involved. This is not unusual where tornadoes have been spawned from supercells.

Tornado Tracks: An Irish summary

It has long been assumed that the most severe tornadoes leave damage tracks that are the longest and the widest. This is not always the case and the evidence from Ireland tells another story.

The length of tornado tracks

A large population of measured tornado tracks has been assembled from site investigations across Ireland. Obviously, such track information can only be determined from site investigations, since reports from the public do not contain this type of information. More than 110 tornado tracks have been identified and measured. These provide a basis for significant conclusions about the tornado footprint in Ireland.

By far the majority of tornado tracks are less than 3 km long (Figure 8.7). However, it is noteworthy that 27 per cent are over 5 km, while 5 per cent are over 10 km. The longest was the remarkable Ardmore/Tolans Point tornado on 31 December 2006.[11] This tracked 30.2 km, mostly across the countryside of County Antrim (Figure 8.8). Its first damage was actually in County Armagh, near Ardmore, before crossing part of Lough Neagh to Tolans Point in County Antrim.

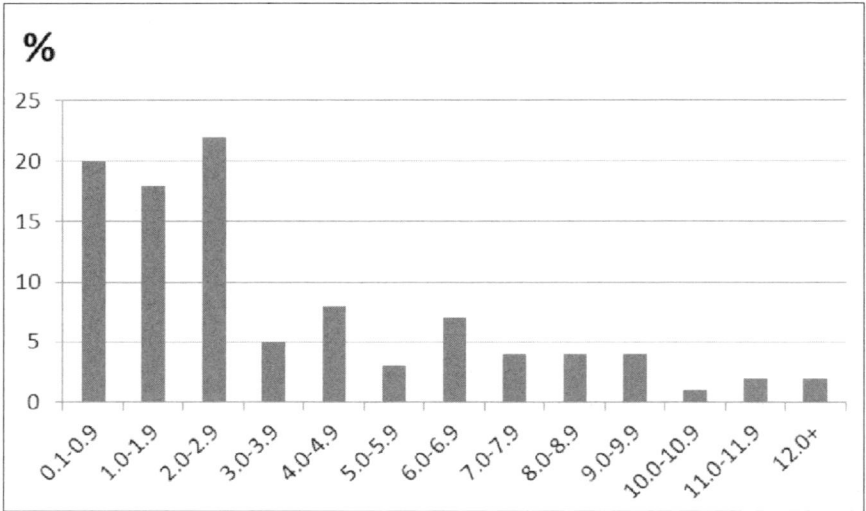

8.7 Tornado track length (km) frequencies by percentage.

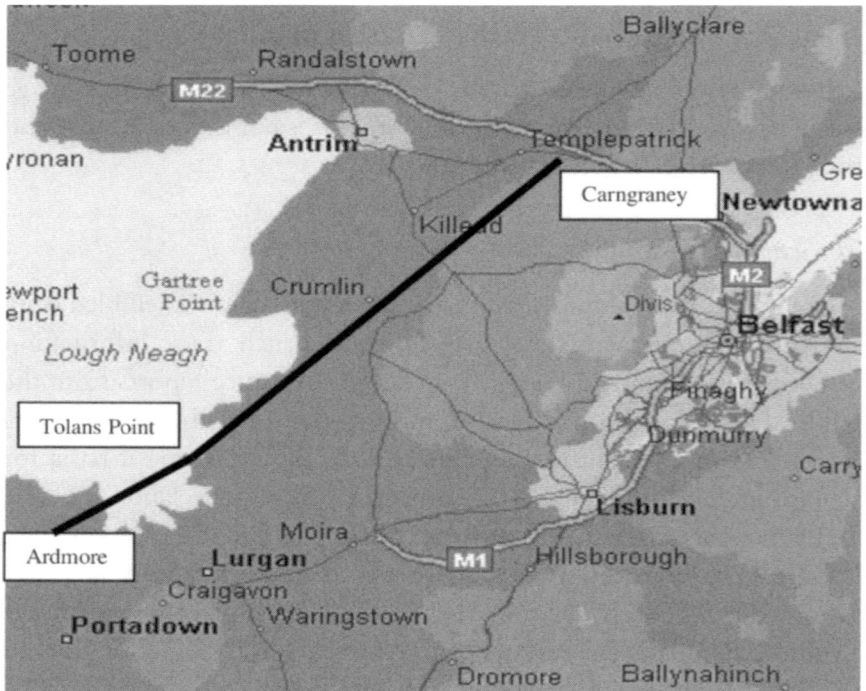

8.8 The Ardmore–Tolans Point tornado track.

It was observed as a waterspout as it did so. Tolans Point is a small peninsula extending into the lough. The damage the tornado caused there showed that it was gaining in intensity as it crossed another small bay, passing Derrymore Point to make land again at Selsham. Being Sunday lunchtime it was seen by scores of people, despite the rain that accompanied it some of the time.

As it tracked across the countryside it was accompanied by rain that was locally so torrential that it produced mudslides on very slight slopes. There were scores of reports of large trees being uprooted or snapped like matchsticks. People were amazed how 'trees were tossed down like a deck of cards', buildings lost their roofs and cars were lifted. There were remarkable debris effects such as heavy items being lifted and seen spinning upwards into the air before becoming embedded at steep angles into fields and house roofs. Debris was carried distances over belts of trees, some of which were themselves ripped apart. The track was a single, continuous one, remarkably straight from a south-westerly direction. It widened twice, once to between 220 and 250 m west of the village of Glanavy, then again at Nutts Corner to 150 m, where it reached its maximum intensity of T5. However, it was consistently 70 to 75 m wide for most of its length.

The start and end points of this tornado were well marked. At first, near Derrycor and towards the lough, it was relatively weak (T0) and narrow, damaging guttering, gardens, hedging and small trees along a path that was a mere 10 to 20 m wide at most. But it gained momentum as it crossed the lough, where it was seen as a growing waterspout by many. At its termination it was still about 50 m wide as it dissipated rather quickly near Carngraney. During the field investigation it quickly became apparent that this was an event of particular significance. Therefore, in the detailed recording and analysis of the tornado's impact on the ground, consideration was given to the possibility that the damage track did not necessarily coincide with the full surface path of the wind vortex and particular care was taken to look for minor effects at each end of the track.

It has become apparent in a number of cases that because the start and finish points of the track are determined by damage features, a shorter path length may be allocated to a tornado than was in fact the case.

This is suspected as being particularly true in Ireland where building designs have long taken some account of the often severe windiness of the climate, so that damage thresholds may be higher than in some other countries. The adaptation of plants to strong winds (through their rooting mechanisms, etc.) is also a factor. This applies not only at the beginning and end of the track, but also throughout its length.

The width of tornado tracks

The width of the damage track is an important indicator of the width of the tornado that caused it. In a general sense this is a valid association. In very precise work there are problems with this. In most cases it is simply impracticable to be able to gather width data other than the maximum width and, with lesser accuracy, an average width.

Most tornadoes in Ireland have track widths that reach less than 30 m. But some are much wider than this (Figure 8.9). A significant 25 per cent are over 60 m wide and 13 per cent are over 100 m wide.

The widest tornado track was measured in Togher, County Meath in April 2006 (Figure 8.10). There, a tornado created swathe of damage 460 metres wide, which was particularly evident in the destruction of trees, some relatively large, along field boundaries. However, the width along most of the 3.2 km track was mostly 30 to 40 m. The transitions before

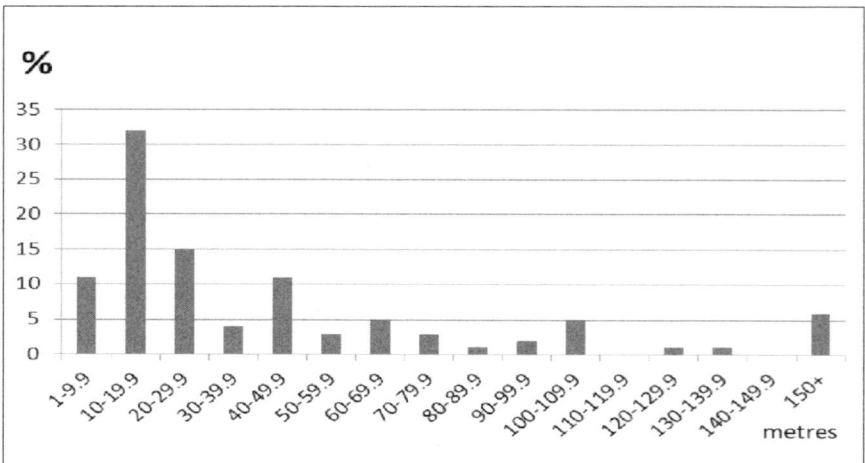

8.9 The frequency of tornado path widths (m), by percentage.

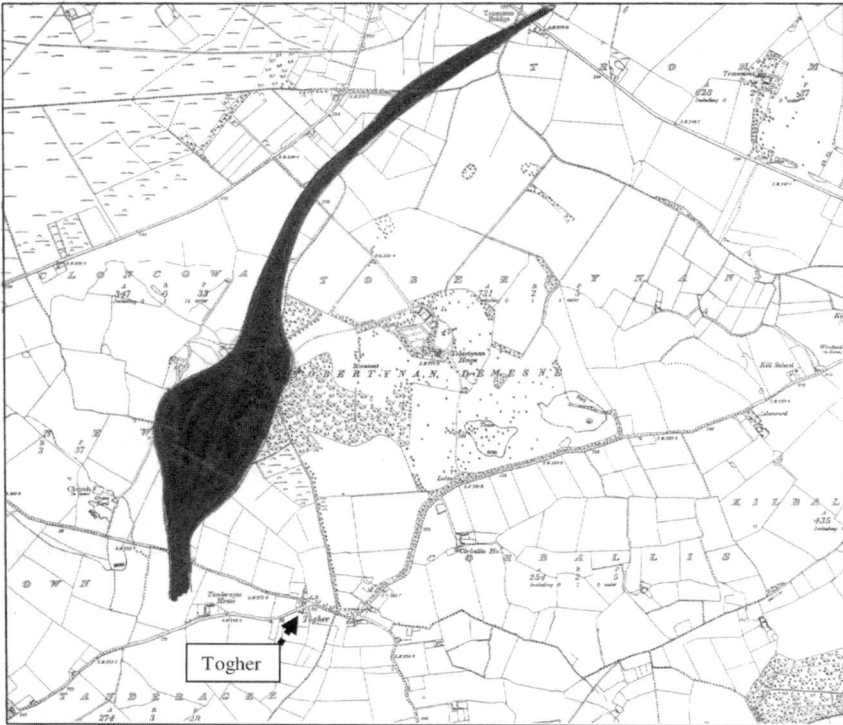

8.10 The track of the Togher tornado.

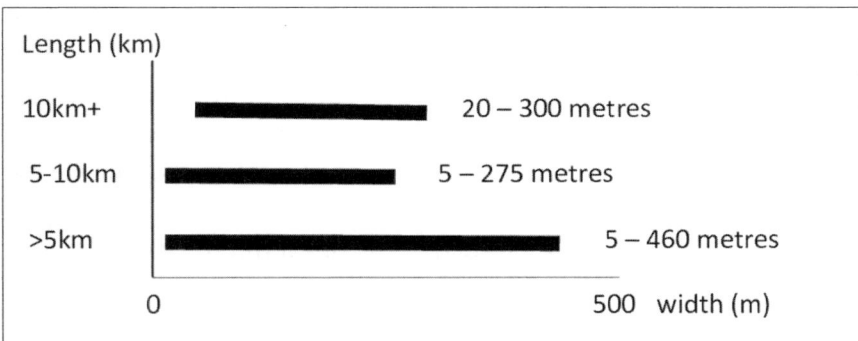

8.11 How tornado path width varies with tornado path length.

and after the widest segments of the track were sudden, increasing from 30 to 400 m over a distance of approximately 300 m. Beyond the widest part of the track, the change was also sudden, quickly narrowing to 45 to 50 m as the tornado continued to forge a track across the countryside.

Individual cases are sometimes used to show that the longest tornadoes tend also to have the widest damage tracks. The Ardmore/ Tolans Point tornado is a good example of this, with the longest and second widest track in all the cases examined. However, a comprehensive examination of all the cases shows that this relationship is not strong. Long tracks frequently have widths that are relatively slight, while the widest tracks are sometimes relatively short, as demonstrated in Figure 8.11 for 110 tornadoes in Ireland for which the relevant data was available from site investigations.

The narrowest tornadoes can be quite striking phenomena. A small number have been less than 10 m wide yet several kilometres in length. Such was the tornado that became a waterspout at Belmullet in County Mayo during April 2001. It is very rare for a tornado to be close to one of Met Éireann's meteorological stations, but this caught one a glancing blow. One of the staff, Brian O'Shea, was on the roof and saw the tornado develop from its parent cloud, and watched it approach the station. As he took cover it came within a few metres of the station. It was so narrow, up to a mere 7 m wide, that it was able to pass between the homes around the station, as it continued on its way. It then tracked a full 8 km before it terminated on Blacksod Bay, having been observed as a waterspout by a number of people along the shoreline.

Estimates of width from visual observations of eyewitnesses, professional or otherwise, may be underestimates of the true diameter of the tornado, because what is seen is determined by variations in the low-level water vapour that can be condensed into moisture droplets, and the result can be the invisibility of the wider rotation of the vortex. It should also be noted that track widths and whirlwind widths may not coincide since damage is not caused until critical threshold wind speeds occur that initiate damage to structures that are present.

Tornado intensity and track dimensions

It is too simplistic to say that stronger tornadoes produce longer and wider tracks. It leads to expectations that cannot be sustained. It is often assumed that the longer or wider the tornado track, so the stronger the tornado must have been.

At first, the generalisation seems to be valid since Irish cases show that track lengths greater than 3 km were predominantly created by tornadoes of T2 or greater. Such a conclusion, derived from so many case studies, disguises a complexity that demonstrates the care with which such conclusions should be applied. The exceptions have been few, for only 8 per cent of tracks of 3 km or longer were created by relatively weak tornadoes (T0 and T1). Nevertheless, they did occur. But it is also the case that 50 per cent of tornadoes with track lengths of less than 3 kms were of T2 intensity or higher. So, stronger tornadoes are just as likely to produce short tracks as weak tornadoes.

Similar caution is necessary in making generalisations about track widths. Trackwidths of 50 m or more were predominantly products of tornadoes with intensities of T2 or higher. But again, there were exceptions to this generalisation. Tornado track widths wider than 50 m were created by the weakest tornadoes (T0 and T1) in 10 per cent of cases. But it is also the case that 60 per cent of tornadoes less than 50 m wide were of T2 intensity or higher. Thus, stronger tornadoes are just as likely to produce narrow tracks as weak tornadoes.

So, the popular, largely intuitive, association of large diameter tornadoes moving swiftly across the landscape with being the most damaging and threatening has not been validated by on-site research investigations in Ireland. A growing awareness of this in other countries as well has led to extensive laboratory simulation tests that found many cases where smaller diameter tornadoes produced larger peak forces, being tighter and more organised.[12] These simulations also showed that higher forward speeds of the vortex produced lower peak forces, especially when these were associated with vortex tilting.

Gathering the Data

The investigation of damage, debris and other effects of a suspected tornado is essential if a reliable database for these events is to be built up. Reliance on reports alone, without on-site visits and assessment, cannot be an effective substitute. In a post-event site investigation, there are two sources of information that can be sought. The first is an impact survey along the supposed track of the tornado. This will be mostly of damage effects. But a second source, yielding quite different information, are eyewitness accounts.

In order to maximise the utility of the resulting data, it is important to be systematic and rigorous in gathering these. Only then will the data be useful for comparing events and building up an understanding of what has occurred.

The following details are provided to explain how site investigations are carried out. Site investigations provide the data upon which the study of tornadoes in Ireland has relied to date. These details will also assist the growing number of weather enthusiasts and students who are beginning to take part in site investigations in Ireland and who lack a handbook of field methods.

Stage 1: Finding the tornado

This may appear to be a remarkable thing to have to do. But, on occasion, it has proved to be particularly difficult. Tornadoes have not been widely reported in Ireland, even when they have occurred. In very recent years there has been a wider interest in them and a partial acceptance of their reality. So, traditional inhibitions in reporting tornadoes are being overcome and a greater readiness to make a report of a possible event has developed.

In most cases, the best place to get started is with reports of severe, highly localised, wind damage. For general weather enthusiasts there are three main ways to access such information: web-based weather discussion groups, the local media and membership of voluntary research organisations (such as TORRO). In addition to using these sources, some weather enthusiasts in Ireland anticipate possible events by monitoring the weather, keeping an observant eye on what is going on in the sky above and networking with others inclined to do the same. In this way it is possible to assess the likelihood of a tornado-spawning storm environment. Occasionally this leads to storm-chasing activity.

In practice, most events are reported by a single individual. However, in a limited number of cases there will be more, particularly when the event occurs near an urban area. When these reports are made to local radio stations or local newspapers some of the information can be lost when no contact information is retained. So whether an event will reach the site investigation stage will depend on personal initiatives within media organisations rather than particular policies guaranteeing follow-up and verification. Sadly, some important reports have been lost to the Irish record because of the failure to do this. However, many of those who have made initial reports have also been very helpful in providing local briefings and other assistance to investigators visiting the sites involved. Our knowledge of Ireland's tornado story owes a debt of gratitude to them.

Since many tornadoes in Ireland occur in relatively remote areas, damage sites can be difficult to find. Indeed, many cases have occurred where there has been little or no local knowledge of their precise location, especially near forested areas or boglands. In addition, reported tornado condensation funnels observed from a distance have sometimes been difficult to locate. Then, a search strategy, using ground transects to intersect with likely tracks suggested by any available local information, has been used to good effect.

Storm damage sites are usually visited as soon as possible following an event. In urban areas this is usually within twenty-four or forty-eight hours. Such a response is necessary because impacts and eyewitnesses disappear quickly as repairs are made and people become more difficult to trace. But in remoter rural Ireland it has been found that the most widespread knowledge of an event may be some days afterwards, when local knowledge and experience have been passed around. Then it is much easier to tap into the pool of information that has been built up and about which there is still an active knowledge. In addition, much damage may remain evident for a long time, as the necessity for clearing is much less.

But the tornado track is a damaged and unstable environment. It still poses a hazard, especially in rural areas, where there are few people around, trees still collapse suddenly and tiles fall from damaged farm outbuildings days after the event, especially if there is a strong breeze at the time of the investigation. This is more obvious in an urban environment with rescue services on the scene.

Stage 2: Detailed mapping

The investigation strategy is to follow the tornado track from its start to its end. This may require some rough and muddy walking and scrambling. The site investigator will always need to be prepared for this. A distinction has always been made between two types of evidence that the site investigator seeks to map: damage and debris. Damage occurs where natural or engineered structures are still in their original location but have been degraded in some way due to the impact of the tornado. This is usually in the relatively dramatic form of components of the natural landscape or built structures that have been broken and have sometimes lost a portion of their material. Debris occurs when material is displaced from its original location, having been removed in the first instance by the winds of the tornado. It includes both whole objects as well as fragments. Most of the debris will occur along or adjacent to the track, but some may be carried considerable distances. The debris field may be wider than the tornado's path across an area and should, therefore, be differentiated from the damage on-site investigation maps.

A systematic method for recording damage and debris gives order to a large amount of data and helps in the later interpretation of the event. On a map scaled to show field boundaries, all such features can be marked with a code (number or symbol) and in the map margin or a small notebook (suitably prepared beforehand) the details shown in Table 8.2 recorded.

Table 8.2 Map annotations for field mapping of a tornado track.

> a. What the feature is and whether damaged or deposited.
> b. If it is damage, the compass bearing of its new orientation.
> c. If it is debris, the compass bearing and distance from its original location.
> d. Any significant details (e.g. height at which a tree is snapped off).
> e. An annotated sketch, if necessary, for later additional comment.

There have been many occasions that damage and debris occur in the most unlikely places. It is common for debris to be found in the upper branches of large trees. This can be more than tree debris. It can include iron sheeting, farm implements, feeding troughs, garden furniture, barrels and roofing materials. Lighter debris such as washing, sacking and fencing can be found hanging on telephone and power lines. These provide evidence of the way a tornado travels and can be especially useful when there is a significant break in the ground-level evidence. In urban areas the need for roof repairs, etc., often make these features very conspicuous.

Stage 3: Eyewitness interviews

Eyewitnesses are those that saw the event at first hand. They are very important and they need a considerable amount of encouragement to come forward with their information. In Ireland, it is quite clear that in the past many eyewitnesses have been reluctant to speak of what they have seen because of anxieties about the reception they would get from others. This has been partly a consequence of the strength of prevailing establishment views (until relatively recently) about the Irish climate that had no place for tornadoes. Only a small proportion of eyewitnesses have made any kind of report, and this has led to a considerable loss of information. Each eyewitness is important because each has a slightly different experience of the event. It is true that some understate what they have seen while others may overstate it, but this is not a reason to reject this evidence. The larger the pool of eyewitnesses that contribute their records of the event, the more useful is their information, because a lot of the information can be validated with reference to other eyewitnesses.

From time to time an appeal for eyewitnesses is useful, using posters in libraries, pubs, rural shops, etc. Eyewitnesses are usually those who were outside at the time of an event. It can be expected that this would consist of a wide cross-section of the population, including children and young people. The latter are often overlooked in site investigations but where they have been included they have produced a fund of important information. To identify those who are likely to have some experience of the event, it has been productive to review the range of activities taking place at the time, not just by those at home or perhaps out on a farm – but also those at petrol stations (regular motorists who fill up at the

Table 8.3 Information gathered by eyewitness questionnaires.

a. Time, date and location; all as precise as possible.
b. What was observed of the funnel structure and its parent cloud (sight and sounds), the sequence of its development (with timings), its changing dimensions, and a sketch of these. Initially witnesses are usually unwilling to sketch, but when persuaded to provide one, it has been very valuable.
c. The sequence of weather conditions immediately before, during and after the tornado, with special reference to rainfall (especially intense rainfall), hail and lightning.
d. Knowledge of any records of the event, such as photos and videos, that were made by local people.

same station), working on the roads, at construction sites, or engaged in the recreational activities on offer in the area. Examples are golf clubs, fishing events or even flying clubs, all of which have been good sources of eyewitness information. Questionnaires are used with eyewitnesses in order to standardise the information gathered, which is usually grouped into the information categories shown in Table 8.3.

After an interview it is always important to make additional notes, since much additional material is usually provided in the course of the interview that does not fall strictly into answers to the questions posed in the questionnaire. This has to be done immediately as waiting until later results in some details being lost.

Stage 4: Interpretation

The interpretation draws on a range of information that is brought together. In Ireland, some key information that would be available in relatively high technology research environments such as exist in the USA is not available (e.g. Doppler radar, aerial surveys, etc.) and is not considered here. However, there still remains a considerable variety of information sources available to interpret the event. These include the surveyed damage and debris and their mapped distributions, eyewitness records, local weather details with wider regional meteorological data and, finally, any photographic evidence.

Table 8.4 Data for the meteorological environment of each tornado event.

> a. Radar images show cells of high rainfall intensity as well as the development of the storm cells involved.
> b. Sferics to show thunderstorm activity.
> c. Upper air charts show stability and differences in wind speed and wind direction through the atmosphere (indicating wind shear).
> d. Surface synoptic conditions to identify significant atmospheric pressure features.
> e. Surface meteorological observations from meteorological stations in the region affected (the synoptic, climatological and rainfall stations of Met Éireann) as well as volunteer stations that have been quality assessed.

Meteorological data and remote sensing images provide important information about the meteorological environment at the time. This wider context is essential for interpreting the event itself. These data will include those listed in Table 8.4.

Such information has helped to define the atmospheric environment when a tornado is reported to have happened. However, often the data are for areas distant from the event site and their resolution can be relatively coarse. So, in themselves such data have never been definitive. Much more small-scale information is necessary to monitor the scale of processes that produce tornadoes in Ireland if we are to understand them more fully.

The primary task of the site investigation is to find out whether or not a tornado has occurred. Reports of possible tornado events normally focus on sudden stormy conditions that produce a lot of damage over a small area. But other conditions produce similar damage, so each one of these other alternatives needs to be ruled out. In particular, small areas of damage can be created by downbursts. The downburst concept has only been recognised since the mid-1970s. They are strong downdrafts, often produced by thunderstorms that frequently cause considerable damage on the ground. If they produce damage swathes of less than 4 km they are known as microbursts, and if greater than that as macrobursts.[13] So they are of a similar order of magnitude as a tornado. However, they produce an almost fan-like pattern of damage that spreads out as the descending air meets the ground. In contrast, the tornado draws air into

Table 8.5 Contrasts in the damage patterns of the tornado and straight-line winds.

Damage characteristics of a tornado:
Damage path is long and narrow
Damage gradient (transition from damaged to non-damaged areas) is sharp
Damage vectors (downed trees or crop stalks) may show a convergent pattern
(into the vortex as it moves)
Damage may appear to be 'chopped up' or chaotic
Damage vectors may show swirls or vortex marks or herringbone pattern
Missing items as well as debris lifted to significant heights to be deposited
beyond buildings and other barriers

Contrast with damage patterns of straight-line winds
Damage path is short and broad
Damage gradient is low
Damage vectors show a divergent pattern
Damage appears to be laid out neatly
Damage vectors may produce a fan-like or star-burst pattern if caused by a
microburst
Debris is driven along the ground surface, some accumulating at barriers

---

the vortex and produces a different pattern of damage. This is just one way to differentiate between the tornado and extreme gusts or straight-line winds (Table 8.5).

When the sequence of weather events is put together from eyewitness accounts, surface observations and remotely sensed products, helpful inferences can be made about what has happened. A tornado is likely to occur before large hail or other severe winds if it is a non-supercell tornado at a convergence boundary. Another situation where this sequence occurs is at the developing or early mature stages of a thunderstorm. Then the storm consists only of an updraft.

Undue weight of proof is sometimes given to photographic evidence. But photographic evidence is not in itself conclusive that a tornado has occurred. If it is available it is important, but normally it cannot stand alone. In the same way other types of evidence require verification. Indeed, there is a growing awareness of the necessity to guard against fake photographic images. The technology to do this is readily available and will become increasingly sophisticated. The diversity of evidence usually available from the site investigation and other information is a significant help in guarding against this.

The evidence from eyewitness accounts is ranked variously in different countries, especially those with developed reporting infrastructures. In those, more professional and trained personnel on the ground may be available to go where conditions are favourable for tornado development and may be at hand to make trained observations. But in Ireland this is not the case. For gathering information about tornadic events in Ireland, the eyewitness account is the major source of information, so it always needs to be sought and then verified on site.

Second to seeing a tornado for real and being able to watch it evolve in its awesome magnificence, a visit to a site struck by a tornado is a memorable experience. Placing order upon the apparent chaos that results from a tornado, and understanding something of the power and energy that carved a path through the countryside, produces many a surprise and is an investigative challenge of the first order.

Site investigations that have assessed track dimensions and tornado intensities have revealed that over the range of tornadoes that have occurred in Ireland, the track length, track width and tornado intensity have no simple relationship. Even with the same tornado intensity, these relationships vary from event to event. This weakens many of the assumptions that are often made about them. As a result, it is always necessary to engage in site investigations with a very open mind. In addition, preliminary comments can never be definitive, even though these are often sought; they always require caution. Every footprint is different.

CHAPTER NINE

Trails of Destruction

Like a claw gauging the earth
It lands with destructive intent
Nothing too small or too large
Resists its grasp or makes it relent.

A we and amazement are often the experience of the beholder of a tornado at a distance. But if the encounter is close, a tornado's destructive power is all too apparent. The more familiar thunder and lightning of dramatic storms in Ireland, which may have preceded the growing funnel, fade into insignificance. Tornadoes mean destruction. Everyone knows that!

But all will not be lost. Detailed site investigations that have mapped and recorded the destructive results of the impact have shown great variation in the damage so inflicted. Places do not get wiped out and everything in a tornado's path does not get swept away. There are subtle interactions between the vortex and the structures it hits as well as between the vortex and the landscapes it crosses.

Structures at Risk When a Tornado Strikes

The impact of a tornado on everything it touches is highly variable. Within and around it are intense convergent horizontal winds. Within its core are strong vertical winds as well. Both large and small structures are at risk from these, as they combine in different ways, whether the vortex diameter is large or small. For example, in small, but intense, vortices, the converging winds on reaching the vortex core often drive upwards at a proportionately much greater speed relative to the convergence speed. Consequently a building or other structure caught by a small tornado can be ripped apart vertically more readily than by a tornado with a much larger diameter.

Explosive effects

A particular case of this occurred at Brow Head on the County Cork coast on 25 November 2000. An unoccupied holiday home had part of its roof ripped off, including its tiles, slates and timbers, while around its sides all the doors, windows and walls remained intact. Through the roof were lifted bunk beds (ripped from the walls), mattresses, other furniture, the TV and numerous other items. These were carried short distances away from the building; the beds and TV into the garden (Figure 9.1) while many of the 61 x 61 cm Welsh slates were found three fields away. The house had stood for forty years and had been re-slated only three years before, using a double layer of batons and felting, brass screws being preferred to nails for added strength.

The threat of damage that has frequently been described as explosive has also received some attention by researchers. It was once thought that the intense low pressure within tornadoes caused buildings to explode. This was based on the idea that the barometric pressure in the building became significantly greater than outside. It would require that the building remained firmly sealed, an assumption that has been found to be wanting. Rather, building damage usually initiates from

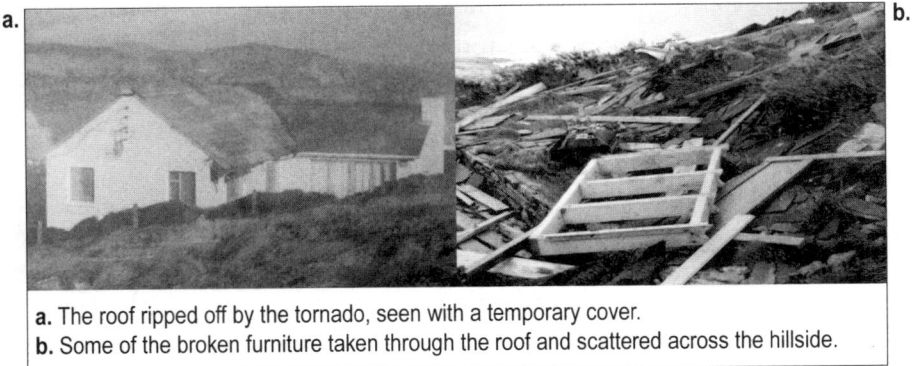

a. The roof ripped off by the tornado, seen with a temporary cover.
b. Some of the broken furniture taken through the roof and scattered across the hillside.

9.1 Impacts of the tornado at Brow Head.

wind pressure on the outside breaching the building, typically through broken windows or doors. Often there is plenty of evidence for this from materials originating from outside being found inside the building, such as mud, glass and wood. Openings on the windward side of the building then increase the internal wind pressure and cause additional uplift beneath the roof.

Nevertheless, there are significant pressure differences that do develop. These are mostly due to the high-speed winds of the tornado going over a building, rather than the lower pressure in the core of the tornado.[1] The forces generated in this way have been found to be two to three times greater than those generated by the lower pressure of the vortex core.[2] As the wind sweeps over the roof, it creates an upward lifting force. The consequences depend on the weight of the roof and how securely it is anchored.

The first record of such an explosive effect caused by a tornado in Ireland was the Limerick tornado of 5 October 1851.[3] As a result of his local enquiries, Dr Griffin of Limerick (see Chapter 4), in his report to the RIA, highlighted two instances where large windows in an otherwise closed and sealed business office shattered 'by a force acting from within'. In light of the discussion above, it should be noted that Dr Griffin appears to have been careful to establish that the buildings were well secured. In the office of Mr Gleeson the window frame was displaced outwards as well as the glass shattered in the same direction. In the case of a property belonging to a Mr Hogg, the damage included 'some windows ... had several panes broken, yet not a trace of broken glass could be found anywhere, neither in the room, on the window-sill, nor in the area below, nor on the flags around it. Some towels also, and a sheet, carried out of one of these windows could never be found afterwards'. Such effects have rarely been recorded in modern accounts of Irish tornadoes.

Building structures

The most important built structure in the Irish landscape is the family home. In many parts of rural Ireland, there has been a history of building houses to withstand severe winds. Now, in modern Ireland, regulations have set the design wind speeds for houses at a relatively high level. They require designs for a gust speed of 44 m/sec, but there are some design requirements of gust speeds to 47 m/sec.[4]

Chapter 6 has shown that by far the majority of tornadoes in Ireland can be described as light, mild or moderate (i.e. up to T2), with estimated winds of up to 41 m/sec. Despite this description, they are very serious events with serious destructive potential. But a significant number do occur that surpass this. They have produced winds that caused major structural damage to homes and similar buildings, all built to the required standards. However, with tornadoes of T3 intensity or greater, the severest damage effects are patchy within the tornado track. This is because only a small portion of the tornado's path actually experiences the highest wind speeds. These occur immediately around the core of the vortex. Further out, but still within the vortex, it is clear from the damage maps that wind speeds are less. So, fewer houses (and similar buildings) within this outer zone have been found to be significantly damaged and site investigations have found many to be relatively unscathed.

The most common damage is to the roof of a building. The wind speed criteria for building standards requires that all roof tiles should be pinned. However, site investigations in Ireland have found roof damage involving the loss of roof tiles where many were unsecured. But it would take an impractical regime of inspection to ensure adequate standards are always maintained. So in Ireland, many homes are more vulnerable to wind damage to their roofs than would be the case if the standards were always met.

A further factor applies with regard to the vulnerability of homes to tornado winds. Those critical wind speeds recognised as causing such damage have been identified and tested by models using straight-line wind assumptions. But tornadoes are not straight-line winds. Their winds are extremely turbulent. They have a strong vertical dimension and produce high suction where there are negative pressures. In addition, they have radial accelerations around the vortex of the tornado. As a result, equivalent damage is likely to occur at much lower wind speeds than in straight-line winds. So, the front part of the tornado hits the building first and suddenly imposes an intense positive pressure. When, perhaps less than a minute later, the rear portion of the tornado encounters the building, the wind blows from the opposite direction, contrary to the direction in which the vortex is travelling. Modelling studies have shown that this produces negative pressures and upward forces, adding to the overall damage potential.[5]

A very common concern has been whether homes are safe places to be in. The biggest threat to the structure of a house comes from the tornadic winds penetrating the building through broken windows and doors. Evidence from within some buildings thus affected has shown fragments of objects, as well as soil and bits of plants, have entered from the windward side. Indeed, there have been cases where flying debris has created an opening for the wind to enter. Even winds as low as 51 km/h (32 mph) enable small missiles to penetrate walls. If this happens the internal pressure may increase and cause an upward pressure on the roof structure. But there is little an occupant can do about this. In Ireland, storms are very frequent, and it is normal for doors and windows to be closed quickly when such storms threaten.

However, in Ireland there has been a quite a dramatic switch to timber frame housing. Ireland had been lagging behind this international practice for many years. In 1992 only 5 per cent of newly built houses were constructed in this way. But by 2009, at the height of the house-building boom, this passed 30 per cent. Although there are many advantages in having a timber frame home, they are more vulnerable in windstorms, as international experience has discovered. Structural connections within the building may be weakened, or possibly fail, especially the roof joist/ top plate and the rafter/top plate connections. These are even more vulnerable to uplift forcing if hit by a tornado, compared with being hit by straight-line storm winds.

Large span roof structures

Large span structures such as auditoriums, gymnasiums, swimming pools, light industrial structures and large new farm out-buildings are examples of the increasingly wide range of new elements that have become part of our urban and rural built environment. These are particularly vulnerable in high winds owing to their large roof expanse upon which wind forces act and the distance between roof and

9.2 Twisted structural damage due to a tornado at Coolrain.

supporting walls. The effect of high velocity wind flows across these roof areas is proportionately greater than over smaller house roofs. Also, the opportunities for wind penetration at the side are greater. As a result, it is important to avoid these when seeking shelter during any severe windstorms, but particularly a tornado.

In rural Ireland such structures multiplied as land from the great estates was redistributed and grant aid became available to construct hay barns from the early twentieth century, but particularly from January 1948 with the Farm Buildings Scheme. Another spurt in the dimensions of farm buildings came in the 1970s after Ireland joined the (then) European Economic Community (EEC), with more intensive farming requiring larger milking parlours, large pig production units, winter housing and ever larger hay barns. The latter now have very high interiors as well as much wider spans. They are often subdivided to house livestock and store large modern tractors and farm machinery. Initially open sided, they often have concrete or sheet metal sides added. Thus, the potential for significant damage by tornadoes in the rural landscape has risen at an increasing rate over recent decades.

Elements of these structures, as well as other smaller ones, may become twisted during the passage of a tornado. Such twisted damage is sometimes taken as information about the rotating wind field of the tornado, but this may not be the case. When a tornado has a diameter greater than the width of a building, it may act as a straight-line wind in passing over it. In many smaller tornadoes this is not so. However, in Ireland it has been found that the cause of twisted building damage is often where a steel frame anchors a building to its concrete base so that the winds of a tornado engulf the upper parts of the structure and twist the frame's steel strut anchored in its concrete base, exerting a massive force upon it to do so. This can be seen clearly in the Coolrain event in County Laois, where a steel girder of a barn was anchored into a concrete base (Figure 9.2).

Travel and Transport over Land and Water

The transport network across the country consists of road, rail and water networks together with a variety of air traffic routes. Although these have a limited amount of important structural elements, the main threat from tornadoes is to those who are using these networks when a tornado is moving across the countryside. As they travel along the routes travellers will normally have a very limited view of the sky. As a result, those encountering a tornado will have been largely unaware of its approach.

Vehicles caught by tornadoes

In May 2003, a school principal driving towards his school sports ground outside Rathangan, County Kildare, drove into a tornado. Without any warning it burst through the trees lining the narrow lane into which he had turned.[6] Conditions were stormy and the tall thick trees and flat landscape ensured that his line of sight was totally obscured. He was thrown about in his car as it rocked up and down violently by half a metre or so on each side. At the same time it was bombarded with debris from trees as well as grit from the road. Although the car was damaged and its windows broken, it was not lifted off the ground. He managed to escape without serious injury.

But a car was lifted off the ground during a much more severe event near the RTÉ (Radió Teilifís Éireann) mast on Kippure Mountain, County Wicklow in July 1983. A team of engineers was working on the mast when a thunderstorm began. The storm produced a tornado which headed towards the group. Having lifted piles of turf into the air, it then lifted one of the cars with the driver inside, raising it about 10 m and carrying it some 60 m into the adjacent bog before dumping it there. Fortunately, the driver received no serious injury, although the car was badly damaged. Likewise, in Belfast in September 1982, a tornado lifted a truck containing 2 tonnes of building material 2 m above the ground and carried 7 m across a road.

Vehicles can be deadly places to be when a tornado strikes. Staying in a car in the vicinity of a tornado is not a good idea! People in vehicles hit by tornadoes in Ireland are relatively few, but a significant number have seen a tornado while travelling on Irish roads. So unexpected has this been that it can take some time for people to accept the phenomenon as real and to make an appropriate response. In some cases this delay can put them into a more vulnerable position. It is always best to drive away from the tornado, preferably perpendicular to, and behind, its track. Roads in Ireland generally change direction all the time so these may not be safe options. The alternative is to pull over and evacuate into appropriate shelter. The most common reaction in Ireland has been either to try to outrun the tornado or to get closer to grab a photo! Fortunately, most tornadoes have a relatively short life and the road network is relatively sparse, so being hit by one while in a vehicle is uncommon.

Laboratory simulations have provided no final answer as to why one thing gets picked up by a tornado and another does not. These simulations have shown that if a vehicle is heavier in one spot (e.g. a heavy engine in the front, or a long empty shell in the rear) then it is easier for a tornado to flip and fling it. If there is an imbalance, one end will get picked up off the ground. Where trucks have been picked up (although no Irish cases have occurred), they have been mostly empty. Also, the more surface a vehicle presents to the wind, the more easily they are lifted and carried from the road. Large trucks, commercial vans and buses are particularly vulnerable.

Everything on wheels, whether stationary or mobile, is vulnerable in the same way. This includes caravans and mobile homes as well as vehicles. Even the frequent practice of lashing these to the ground has not always prevented serious damage, as demonstrated by the Derrynacross tornado, when even thick wire anchorages in a cement base were insufficient to secure a mobile home which was torn from its base, lifted across a lane and dumped.[7]

Transport systems

As Ireland has developed its infrastructure over recent decades to accommodate the requirements of a rapidly developing economy and modernising society, so the various expressions of increased communications and interconnectedness have changed its transport systems in many ways. Air links, rail networks, wider and more direct road routes with long-span bridges sweeping across valleys and estuaries have all replaced former more tortuous, slower routes. But the likelihood of a tornado hit was never in the mind of engineers, planners and designers.

Few people in modern Ireland realise how exposed they may have been to the catastrophic effects of a tornado hit. There have been an increasing number of near misses that could have caused major tragedies. This could not be better illustrated than by the near miss of the Ryanair flight between Luton and Kerry Airport on 16 December 2011. A tornado developed close to Castleisland and tracked some 7 km to the south-east. A detailed site investigation found the continuous track reached close to the village of Scartaglen, thereby crossing the plane's flightpath as it descended into Kerry Airport, just 5 km to the west, at the time when it was due to land (Figure 9.3). The pilot did not see the tornado, nor were the flight control staff in the tower aware of the threat, their line of sight being completely hindered by the poor visibility caused by the surrounding stormy showers. But on the ground there had been numerous eyewitnesses.

The invisibility of the threat in such circumstances is equally present on the ground. The tornado that developed at Straboe, County Laois, on 22 June 2011, crossed the main Dublin rail line less than a minute

9.3 The Castleisland tornado track and the landing approach into Kerry Airport.

after the Cork to Dublin train had passed by. The tornado's damage to homes and other buildings on both sides of the track was considerable, cutting a swathe across the immediate countryside some 40 m wide for 2 to 3 kms. Had it hit the train the results would have been catastrophic.

The road is not a safe place when tornadoes occur. They frequently cross roads, sometimes more than once, becoming a serious hazard for all types of vehicles. It is not unusual for investigations of individual events to reveal a considerable number of near misses, particularly in the wider relatively dense road network in the east of the country as well as other high road-use areas near the major cities.

A relatively new element to this risk are long-span bridges, especially the cable-supported designs. These have become more popular in Ireland as they often shorten journey distances and times dramatically and give a strong positive statement about modernisation to everyone. They range from motorway overpasses to pedestrian footbridges across rivers. They are highly sensitive to wind conditions in a way that their predecessors were not, due to their relative flexibility and line-like structural character. This relates to both the deck of the bridge as well as its supporting structures. In their design stage, modelling the effects of severe winds on their behaviour has become a standard requirement. But it has been noted that the assumptions built into these models are not all appropriate for some extreme events such as tornadoes.[8] In addition, the adequacy of the wind event database required for such testing is far from what is required. The Waterford Bridge across the River Suir is the most recent and the longest such bridge in Ireland. It uses a single main concrete pylon 115 metres high with numerous stay cables from its top to support the span on each of its sides.

Water bodies

All bodies of water in Ireland are vulnerable to the development of waterspouts. They have occurred on inland lakes of all sizes as well as along all coasts. There, promontories, coastal bays and offshore islands appear to contribute significantly to their development compared with areas where such features are not characteristic (see Chapter 5). However, compared with land surfaces, all water bodies have much lower thresholds for producing dangerous conditions, whether to life or to sailing vessels. So they are particularly high-risk environments when other atmospheric conditions are suitable for vortex development. But many tornadoes in Ireland spend part of their life over water and sometimes are referred to as tornadic waterspouts. They tend to be among the most dangerous waterspouts – much more so than those that remain on water surfaces throughout their existence.

Threats on water vary, but even the weakest waterspout is dangerous to water users. Table 9.1 is based mostly on Irish events but, since their number is relatively low, cases in the wider area around Irish waters have been included.

Characteristic experiences are the suddenness of the waterspout appearance, the occurrence of more than one and the violence of the impact. On inland waters and estuaries it is sometimes difficult to avoid them because space can be very limited. Some fast boats have got around them instead of taking the most advised strategy. The latter is to head 90 degrees away from the estimated track direction and bear in mind that

Table 9.1 Waterspout threats on Irish waters.

Vessel type	Direct hit	Near miss
Small leisure boats	All capsize	Break from moorings
Yachts	Majority sink	Minority capsize
Trawlers	A few lifted and dumped	Deluged by spout water
River ferries	Some swept onshore	
Freighters	Rotate 90 degrees or more	Listing – cargo shift
Sea-going ferries	Severe damage on deck	Below deck minor damage
	Some below deck damage	

there may be erratic deviations from this. Other water users have had no option but to retreat ahead of the vortex. In the cases reported this has worked.

It has been noted that the greatest risk to small vessels may well be when a waterspout collapses. When this happens many tons of water may drop in a small area from the sea spray lifted tens of metres upwards by the waterspout's column and from condensation within the length of the funnel. Indeed, it has been noted that a shallow dome of water may build up from the water surface within the lower vortex, which is then suddenly released when the lower part of the vortex disintegrates. Occasionally this has produced a surge of water that can break both large and small boats from their moorings, depending on the size of the event.

Some of these features were dramatically illustrated by the waterspouts on Blessington Lake, County Wicklow, during a regatta in April 1998. The first yacht race started with light variable winds and occasional sunshine. Then the winds began changing. Some of the crews took advantage of this and the lead changed hands, but this bunched the boats together. The crews were aware of the changing wind conditions since they were vital to them. But going round their last gybe they were suddenly confronted with tornadoes forming on the water under storm clouds. One of the crews avoided one of the waterspouts but another suddenly caught up with them and 'threw the boat over like a toy'. Others were drenched and rocked violently in the vicinity. This was the only race of the day because the wind then died down to nothing.

A number of cases have occurred in recent years when waterspouts have lifted small craft from the water and dumped them on to the

9.4 A boat lifted from its moorings to the back of the beach at Broadhaven Bay, County Mayo, August 2005.

adjacent shoreline, as in the Inver tornado at Broadhaven Bay, County Mayo (Figure 9.4). Elsewhere, in County Kerry, two people and an outboard engine on board a fibreglass boat were lifted twenty feet off the water – and survived.

There are few records of direct hits on large vessels in Irish waters. However, even as early as 1834 it was recorded that a waterspout on the Irish Sea hit the packet steamer HM Packet *Thetis* a glancing blow as it partially went over the 260-ton vessel in the early daylight hours. But no serious damage was done. The waterspout was observed initially from the vessel when it was about a mile away, but the *Thetis* still failed to avoid it. However, there have been a number of other cases where waterspouts have been observed and photographed nearby, when evasive action was taken. Some of these have been from ferries while others have been taken from the shore.

Urban and Rural Environments

Different landscape types can have their own impact upon tornadoes. Some are more resistant to the damage and destruction tornadoes can cause. This resistance produces its own effects, bringing modifications to the vortex close to the ground surface. This can sometimes have a significant effect upon a tornado's longevity.

Urban effects

That weak tornadoes tend to be dissipated when they encounter the increased surface roughness of an urban area tends to be supported by the Irish experience.[9] But despite this effect, the concentration of homes as well as commercial and industrial establishments in urban areas makes tornadoes a significant threat to the population, and economic losses may be incurred. Such threats will only grow with continuing and spreading urbanisation in Ireland. The County Wexford tornado of 7 January 1998 is an excellent example of this. It touched down on the slopes of Forth Mountain and tracked eastwards to Wexford town, dissipating as it reached the town centre, just one street away from the harbour. The most significant impacts were in the suburbs, inflicted upon residences and an industrial plant. The tornado's intensity was very average, a site investigation assessing it as a T2 event.

In contrast, the Limerick tornado of 1851, which entered the city from the north-west and crossed the Wellesley Bridge, was a T3 event, and inflicted damage in the centre from Honan's Quay to Denmark Street, Ellen Street, Margaret Street, John Street, Old Clare Street and Pennywell, before continuing eastwards away from the city (Figure 9.5).[10] There have been a considerable number of tornadoes in Ireland that have developed outside towns but have then tracked right across them in a similar way. More recent examples are Mullingar on 30 September 2001 (T2), Athlone on 12 January 2004 (T2) and Kilmallock on 18 December 2013 (T2).

That relatively weak tornadoes may pass through towns and cities without being terminated by the disruptive effects of urban structures does not fit comfortably with the conclusion from studies in the USA that it is moderate to severe tornadoes that are able to do this rather than weaker ones.[11] This is because the small-scale influence of an urban area is largely negligible for the large-scale meteorological environment that produces the more intense events. But the particular details of the urban

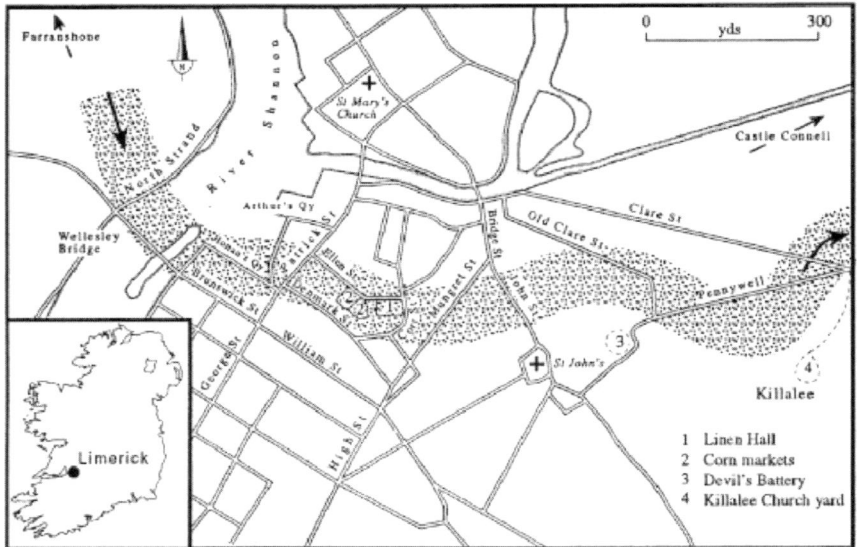

9.5 The track of the Limerick tornado of 1851.

surface may be an important influence in what happens. Investigation of Irish cases suggests that the relative narrowness of terraced town streets may limit the penetration of the vortex to street level, especially when the forward speed of the vortex is significant, and the forward direction is perpendicular to the street alignment. In such cases there tends to be a predominance of roof top and upper level damage. Street-level damage and injury has often come from falling debris.

Urban structures have not prevented tornadoes from developing within the bounds of Irish cities. The Belfast tornado of 26 September 1982 tracked south-south-east–north-north-west for about 1 km from Edlingham Street, right in the central area of the city.[12] The range of damage it caused was such that its intensity was assessed between T3 and T5. The Dublin tornado of 17 October 2012 began in the central suburb of Crumlin and tracked northwards. Fortunately, this was early in the morning when there was little traffic and few people were about, unlike the Dublin tornado of April 1850 that developed within the city and crossed its central areas in mid-afternoon, terrifying Dubliners.[13]

Modelling the effects of urban roughness has suggested that the height of the roughness zone created by urban buildings is important. This limits vortex development as its inner downward flow stays above the zone. Below it the flow becomes much more complicated, tending to break up the typical concentric circles and widening the core of the tornado. Hence, there is a threat of increased losses and wider impacts, even though the severity of a tornado may be modified considerably.

The number of cases of urban tornadoes in Ireland are too few to be able to assess whether the theoretical drivers of tornado development that have sometimes been associated with urban environments apply. While the urban heat island effect can raise the severity of a storm above and downwind of main urban areas it is by no means an established process with Irish cities, even though moderate heat island features are well known. The increased wind shear at low levels that results from the greater roughness of urban surfaces potentially favours tornadogenesis if other conditions are suitable. But there is little direct evidence of this being a major factor in the many Irish cases that have been studied.

Trees and forests

Since Ireland's land area is mostly rural, a tornado's passage is usually marked by tree damage. Although there are large areas of peat and bog where no trees occur at all (except at their boundaries), the history of deforestation has resulted in trees in the present landscape mainly being along field boundaries, and only infrequently clustered in forests and woodlands. Forest covers only 10 per cent of Ireland's land, the second lowest in Europe. So the opportunity for an extensive corridor of damage through woodlands is very limited. But the opportunity increases as afforestation programmes have been promoted, since the alternative to trees is a plant cover upon which a tornado has less visible impact.

The most dramatic damage has been to the largest trees. Along tornado tracks where there have been a wide range of trees (young and old, small and large; all of them healthy), the larger ones have been damaged disproportionately. So, these appear to be more at risk. This not only relates to trunk breakage, but also in uprooting, despite the fact that their rooting system will be more developed than that of younger trees.

The Derrynacross tornado in County Longford on 25 October 2002, to which some reference has already been made, was one of a number in

Ireland where tornado tracks entered forest land and were confronted by a thick wall of trees. This was a severe-intense tornado with a maximum vortex intensity reaching to T4 or T5 (209 to 241 km/h). It formed on the slopes of nearby Corn Hill and descended across farmland to the now forested boglands of Derrynacross. It had already caused serious damage and its intensity had by no means diminished by the time it reached the forest. The forest was a conifer plantation in which the trees were close and the canopy was full. Its mean upper canopy height, measured by the trees that were uprooted, averaged 23 m. Thus, the permeability of such a barrier was slight. The forest edge suffered badly. On immediate impact, scores of trees were uprooted and flattened in directions that reflected the front forward section of the tornado. This swathe was about 75 m wide. The forward movement of the tornado was relatively slow, so its impact was concentrated. However, once the forest edge had been breached, the tree density prevented many trees further into the forest from hitting the ground and they were wedged at angles against adjacent trees. But still many of the tree trunks were snapped off, as well as many branches, some of which were carried away. Tree damage then largely occurred only at successively higher levels in the canopy as the tornado advanced. This resistance to its forward motion was such that the structure of the lower vortex appears to have been disrupted, detaching it from the rest of the tornado. The latter then dissipated.

Tornadoes will track further into more open woodland situations. In ripping off branches, breaking up the ground layer vegetation, uprooting trees and snapping tree trunks, usually at a height of 3 to 6 m, they leave a marked trail. But the effect is similar to the many severe destructive wind events that have impacted Irish woodlands for centuries. These open up the woodland, add to the diversity of species from the surrounding area and contribute to the dynamics of the woodland ecosystem that has developed over time. Such woodland areas now have a relatively small extent, so tornadoes passing through them emerge loaded with debris.

In contrast to the effects upon dense woodlands, the Markethill tornado of County Armagh that occurred on 1 January 2005 passed right through Gosford Forest Park, which is 240 hectares of mixed woodland on the former Gosford Park Demesne. The tornado then continued across the countryside beyond. In the woodland, both medium to large sized trees were uprooted and snapped by the tornado, including several giant trees, among which were important trees from China and one that was the tree girth champion of Britain and Ireland. In addition, large quantities of debris were produced. The spectacular visual consequences of this were described by residents of Mullaghbrack, about a mile to the north of the woodland boundary. They described it vividly:

> We observed the tornado from our picture window, which has views across the demesne. It became really dark as a storm cloud seemed to build up to a thunderburst. As we watched, a grey cloud developed from the base of the cloud. Branches and sticks and other things rose up from the woodland towards this and met it and completed a whole funnel. It came from the forest across the fields towards us. As it approached it had a very distinctive roar that was very different from the sound of a gale. It arrived and damaged the garden, bringing down an old cherry tree and lilac tree and did other minor damage. The house was battered but not seriously damaged.

Old, historic tornado sites in Ireland show that the effects of a tornado have depended on whether there has been a management regime that cleared the felled trees or whether the 'pit and mound' habitats are left alone, together with the tree remnants.[14]

Animals and wildlife

Irish farming has filled the landscape with approximately 8 million cattle, 6 million sheep and up to 200,000 horses. Most of them are kept outside. Intuitively, this would appear to make them extremely vulnerable to severe winds of any kind. But it is remarkable that there have been so few fatalities from tornado events across the country. Hundreds of farms, the majority of them engaged in livestock farming, have been visited during site investigations after a tornado has crossed the farmland. In no case was any large livestock fatality reported. However, on a number of occasions poultry had gone missing, especially when their pens had been damaged or destroyed.

It appears that animals have been safer outside. No detailed research has established this, but anecdotal evidence time and again, both in Ireland and elsewhere, has reported that cattle, sheep and horses are able to disperse into sheltered spots in their pastures if they have the freedom to do so. Many farmers have spoken of them having an instinctive type of weather sense. In contrast, when they are contained in a barn which is hit, they have not been able to avoid the resulting devastation from the impact. In some countries there are many who will deliberately release their animals, including valuable horses, to choose a place where their instincts tell them it is the safest to be.

Wider Hazards: Debris spread, lofting and fallout

One of the most dangerous parts of the tornado is the debris cloud. Most of this is associated with the lower part of the tornado, but in exceptional cases it can extend to considerable heights. The debris field normally commences within the area of the inflow into the tornado funnel at the ground surface, particularly in the area of the inflow jet. The extreme winds there carry debris and relatively loose surface materials (soil, water, woody litter) towards the updraft. There, as it is drawn into the updraft, it is being increasingly concentrated, and this area can have one of the densest concentrations of debris within the storm system as a whole (see Chapter 4). Consequently, it can be the most dangerous part of it.

Debris impacts are directly responsible for a large proportion of the total damage caused in all tornadoes. On impact, debris creates further debris, with the consequence that the total amount of debris will grow dramatically for a considerable distance. While much damaging debris will impact within, or close to, the tornado track itself, much of it will travel well beyond the parent vortex. As a result, a debris field is created that extends for considerable distances.

In Ireland, where eyewitnesses have been able to see the base of the tornado funnel on the ground, the debris has often seemed almost invisible. The cloud that develops around its base appears to consist largely of water droplets, often described as a 'dense swirling mist'. As the tornado develops and tracks across the ground, more of this is added as its circulation intensifies, including debris. This is then lifted in the updrafts to heights that depend on the weight of the debris and the tornado's intensity. Occasionally, relatively light objects have been lifted into the storm cell itself, while many heavy structural materials from farm buildings have been lifted and carried along for only 1 or 2 km before being deposited.

Local debris fields

Site investigations find only a relatively small proportion of the total debris fragments. In most cases it is difficult to establish where particular items originated. But there are circumstances when this can be done, when particular items have a known geographical origin. This is usually large items that are carried relatively short distances. Small debris, such as glass fragments, splinters of wood and roof tiles, can be carried much further and are the most difficult to trace.

In Ireland, the abrupt transition of land use from one type to another has helped to make debris conspicuous during site investigations. Farm building debris carried onto nearby bogs, forest debris carried across a sharp treeline in upland areas and aquatic organisms carried across land surfaces, are all examples. Exceptionally, an isolated home has been hit and its contents scattered, as in the case of the Brow Head tornado quoted earlier. The debris field for this was mapped and included items of all sizes ranging from bunk bed sections and tables to books and paper. The most intense part of this commenced 35 m away from the building and formed a crescent-shaped intense debris zone of up to 150 m wide that ended at a cliff face. From the debris scattered along the shoreline, it was clear that the full debris field extended much further. Had this been in a residential area the impacts would have been severe, being many times wider than the width of the tornado vortex itself.

The smallest debris is probably the most lethal of all. It consists of shards of glass, splintered wood, minute fragments of metal, plastic, pottery and other debris. These can produce a dense cloud that often initially reveals the outer swirl of the vortex to an observer. Then the density decreases as the debris spreads out into the wider airflow. But the debris is still potentially harmful. There have been a number of occasions in Ireland when injuries have been sustained by people exposed to such debris, even though they have been away from the main vortex.

9.6 One of a number of corrugated galvanised iron sheets that went missing during a tornado in County Tyrone, but was found on a distant farm during harvesting some considerable time later.

A common moderately sized debris item is the corrugated iron sheet. Once ripped from their anchor, often from large sheds and farm buildings, such sheets readily loft and travel and are potentially lethal. On 30 September 2006, a tornado hit the relatively new structures of South East Vegetables in County Wexford and ripped off the roof panels which, with large pallets from inside, were lifted some 75 m above the former roof level and deposited 800 m away, over a wide arc of land, which was many times the width of the tornado vortex. The Coolrain tornado of 24 December 1999 dispersed galvanised sheets of 10 to 15 kg across the countryside for 4.1 km in a fan-like arc. The few people out at the time of the storm were able to see these large objects and take avoiding action. In one of the fields the driver of a mechanical digger initially saw the tornado and considered abandoning his vehicle. However, seeing galvanised sheets whirling in the wind 6 m above his head made him decide to stay put inside the cab. But many sheets are not seen and go missing. Some are then discovered on other distant farms at a later date, such as when harvesting occurs (Figure 9.6).

The largest items lofted by a tornado have been quite striking. There have been remarkable cases of whole mobile homes being lifted and carried.[15] These often weigh 8 to 10 tonnes. Some have been rolled over, but a number have been lifted and deposited elsewhere, sometimes upside down or broken up. Fortunately, this has often happened when no one was at home. But the Quirke and O'Dwyer families staying at Dungarvan's Ballyclamper Holiday Park were not so fortunate in June 1998. The Ballyclamper tornado lifted their homes 3 to 4 m, swept them 15 m across the site and dumped them on the ground, hospitalising the nine occupants. Then in July 1999, an occupied mobile home in Carraroe, County Galway was lifted to a height of 15 m, then hit the ground and broke up sufficiently for the injured occupants to escape, but parts of it were subsequently located up to 6.5 km away. These were part of an extensive and very diverse cone-shaped debris field produced by the tornado that was four times as long as the damage track itself and up to fifty times as wide as the vortex.

Long-distance debris fallout

It is one of the strangest things imaginable to be showered suddenly with unfamiliar, unexpected objects from the sky. But throughout Ireland there is a long history of aquatic organisms and other objects falling from the sky and causing amazement and wonder. From the *Annals of the Four Masters*, which records a shower of fish occurring in Tirconnell during 1566, to the present day, such events are traceable as part of the Irish experience.

In recent years there has been a small, but growing, amount of attention internationally given to the question of what happens to debris produced by tornadoes, but which does not fall to the ground within, or close to, the tornado track itself. A debris field is produced by a tornado and its parent storm and extends a long way from the locality of the original tornado. This is caused by the lofting of debris, sometimes to great heights, its subsequent transport by the storm cells that originally generated the tornado, then its eventual distant fallout that will be quite unexpected to any witnesses. Such tornado debris lofting, transport and fallout pose interesting scientific questions and challenges. These are particularly important when assessing the risk if the debris involved is hazardous material such as toxic chemicals or even radioactive waste. It is quite clear that long-distance transport of debris by tornadic storms may pose serious hazards well beyond the direct impact of their high winds.

International research indicates that debris fallout patterns reflect the weight of the object, although when wetted and iced its fall speed may be increased and the distance travelled reduced. Such debris mostly occurs to the left and ahead of the tornado storm track. However, this has not been the case universally as some long-distance debris fallout has been noted to the right of the storm track. Long-distance debris dispersal tends to occur from the parent storm after the tornado has dissipated but the storm cell itself continues to travel, precipitating debris that can come from either its rear flank or forward flank. Much of this depends on how long a storm cell lasts and how far it travels. As long as it continues it has the potential to transport the lofted debris. Internationally, the furthest distance recorded so far is 353 km.[16]

In Ireland, sometimes such falls aroused a curiosity that led to a record being made of the event. Most of such records refer to substances derived originally from an aquatic source, since on landing they stand out in stark contrast to their surrounds and clearly do not 'belong'. Rarely is the explanation immediately apparent. However, organisms that live in water, but fall from the sky in copious numbers, are now often considered to be associated with waterspout activity, although in most instances there is little awareness of a local waterspout event. This has led to the surmise that, on at least some of these occasions, the organisms have been carried considerable distances within a storm cell before being ejected to the ground in a downdraft. Similar falls of non-aquatic organisms are receiving slower acceptance.

The organisms involved are those that live in the surface layers of water. One documented case that occurred in Ireland during the summer of 2000 was of a fall of very small, slim fish, commonly called lesser sand eels (*Ammodytes tobianus*).[17] These live in sandy inshore waters all around Ireland, where they both bury themselves into the sand or swim around in small shoals. On 7 June, in a shower of rain at Ballyconnell, County Sligo, scores of these fish landed on a cottage roof and on the ground around it. This was less than a kilometre from a series of large sandy bays west of Rosses Point extending along the coastline of County

Mayo. Many of the ingredients favouring waterspout development came together during that morning as a small but intense depression arrived off the west coast with a convergent frontal zone: surface vortices, strong low- to medium-level wind shear characteristics and both low and high-level stretching mechanisms. These were sufficient for a relatively intense waterspout, even if of small diameter.

The suction required to lift fish results from the decrease in atmospheric pressure inside the tube of the waterspout. This is normally sharp and occurs suddenly as the stretching mechanisms get to work. The fish are then lifted with water and anything else the water contains, provided there is a sufficient lifting force for the objects concerned. This is rather like the suction of a vacuum cleaner, and is only adequate for a particular range of small objects. The fish that fell at Ballyconnell were not frozen, so they had not been lifted to great heights within the cloud. The freezing level along the west coast on 7 June was at about 1500 m. But no waterspout was observed. However, the weather was described as being very threatening and even the residents of the property affected had taken shelter indoors and saw nothing other than an approaching rainstorm with its dark, low cloud base and curtain of rain. It is likely that in this case the waterspout had a short life span, being swamped by the curtain of rain that washed out the fish after they had been transported only a short distance.

The small village of Dunlavin, in County Wicklow, is much further from the sea. Dunlavin experienced a similar fall of fish on 8 August 1996. The species of the fish that fell on this occasion was not established, so they could have been from either a marine or a freshwater source. The regional airflow was from the south and south-west due to a depression off the north-west coast of Ireland. So if the fish were extracted from their water habitat by a waterspout and carried airborne to Dunlavin from the nearest freshwater source, they would have originated from the River Slaney. This stretches from the coast to barely a kilometre away.[18] However, if they were marine fish, the nearest seacoast to the south is a distance of 120 km. This means that the waters within the Slaney catchment are the most probable source.

There is a further report of numerous small fish of 50 to 75 mm long falling from a storm cell in County Wicklow on 25 October 2002, by campers who emerged from the shelter of their tent to find thirty of them scattered around the immediate vicinity. These could have been from a local freshwater source because, unknown to the campers, one of three tornadoes that occurred that day in Ireland occurred nearby and crossed Derry Water, a tributary of the River Clody, a few kilometres upwind from their location. The identification of the fish species is more definitive of whether the origin is marine or otherwise, so the shower of sticklebacks south of Ballynahinch, County Down, in 1998, clearly originated from a freshwater source.

Such falls from the sky involving single species may be difficult to understand since in any small area of water more than one fish species and other organisms can be expected. However, differences in mass, size, and shape are thought to be part of the answer. Not all falls are like this. Some do have a variety of content. The probability that the most striking deposit of the fallout objects may be preferentially noticed and reported is high.

Not all fish falls are due to waterspouts. On 3 September 2003, the *Western People* had the striking headline 'Flying salmon goes through the roof of Ballina house'. This was accompanied by a photograph of the house owner with the salmon. But the local National Parks and Wildlife Service thought that an osprey probably dropped the fish from its talons while in flight. Similar falls of single organisms are likely to have comparable causes. This makes the examination of the atmospheric conditions over a wide area, both at the synoptic scale and the mesoscale, of critical importance when assessing whether such falls could possibly be caused by a waterspout.

Similar unexpected falls of objects originating from land surfaces have also been recorded in Ireland. These include highly localised showers of berries, nuts, wheat and hay. But they can only be noticed when they fall in populated areas. This makes the likelihood of them being recorded in any way very low. Nevertheless, records of such events span the centuries, and it was shown earlier that they are to be found in early Irish historical records such as the *Annals,* as well as in modern Irish environmental journals, such as the *Irish Naturalists' Journal.*

But probably the most dramatic historical record of long-distance debris fallout is that for the event in AD 847 when the large wooden cross at Slane was carried into the air, as considered in Chapter 2.[19] It was observed being lifted, apparently intact. Then it became fragmented. This happened at a considerable height, well away from the ground. As a result, large items of identifiable debris were scattered. The record of where this debris landed, at Teltown and Fennor, suggests a significant impact (hence 'wonder') for local people on the receiving end, rather than the event having been watched throughout from Slane itself. The information must have been derived from the debris' landing points. From Slane the cross fragment carried to Teltown (Tailltin) covered a distance of 15 km. It was a shorter journey for the other fragment to be carried to Fennor (Finnabhair), just 2 km away. Most tornado tracks in Ireland are less than 5 km long, so this event was somewhat remarkable for its track length.

The significance of such long-distance debris falls is not in their frequency, since the nature of these events makes a consistent chronological record impossible. But the fragmented record that does exist shows that they can be traced far back into Ireland's history. As a result, they are an important additional strand of evidence in establishing the case for tornadoes having been part of the Irish climate for many centuries. Indeed, the scientific awareness of these Irish events has been such that some of them have been quoted internationally since the nineteenth century.

CHAPTER TEN

Popular Responses over Past Centuries

There's no need to fear the wind if your haystacks are tied down.

(Old Irish Proverb)

In Ireland's windswept environment there are various ways of reducing the risks associated with the severest winds. These mitigating measures have lessened the vulnerability of the population to such extreme conditions. However, tornadoes are not ordinary extreme winds, so the vulnerability has persisted. Vulnerability relates to the types of impact that may occur as the weather system passes. It goes beyond the very obvious physical injury an individual may suffer to include the many particular fears and anxieties related to a person's perception of what is happening. The possibility of serious injuries and fatalities are the ultimate expression of vulnerability when tornado weather conditions threaten. In Ireland, such effects and their associated anxieties can be traced back over many centuries. But many of the historical accounts are complex because over hundreds of years all sorts of imagery became incorporated into them and the focus on the vulnerability of people was at best very mixed.

The *Sidhe Gaoithe*: How fear generated respect

For centuries whirlwinds of many kinds were traditionally referred to as *sidhe gaoithe*, as described in Chapter 2. In these earlier years there are various stories of death by a *sidhe gaoithe*. Fatal encounters have always been a real fear. One such is recorded by Lady Wilde (1821–96), who made a significant contribution to the collection of Irish folklore. This was done to contribute to an effort that was being made into understanding the traditional perspectives of the world that characterised the minds of the peasantry at that time.[1] She provides a moving description of a young woman struck down and killed by a faerie blast, with brief details of the debris swirl in the lower part of the vortex. She records, 'a fairy blast had passed over the field carrying a cloud of dust and stones with it, and there could be no doubt but that the fairies had rushed by in the cloud and struck the woman dead as they passed'. The catastrophic consequences of events recorded in accounts such as this strongly suggest the passing of a tornado rather than a small, limited surface-based vortex.

Oscar Wilde (the son of Lady Wilde) carried out extensive fieldwork among the Irish peasantry for similar reasons.[2] He recorded an incident involving numerous faeries in a *sidhe gaoithe*. He states, 'An old man told me he saw them fight once; they tore the thatch off a house in the midst of it all. Had anyone been near they would merely have seen a great wind whirling everything into the air as it passed'. He found that across Ireland there was a significant memory of such events. He recorded some of these. They included an event during 1853 in Kilkenny which resulted in a house being destroyed and a young woman injured by a faery blast. He learnt that in the following year, 1854, there was a *shee gheea* (a frequent rendering of *sidhe gaoithe*) that struck Ballinhassig in County Cork, causing damage. In 1855, a violent whirlwind described to him as a 'fairies frolic' hit Quin in County Clare with very serious consequences. It swept one family from their home, and they were described as having landed up in the ground of the local cemetery. In 1886, it was the turn of Roscommon, when a faery blast tore through it with damage that received the comment that 'cannon balls could not have done greater damage'.[3]

On water, similar terrible events occurred. The storm of 26 April 1894 off the south coast, which included a *sheegwee* (another rendering of *sidhe gaoithe*) ploughing through part of the fishing fleet resulted in many fishing boats being sunk and many lives lost. The coastal community was devastated by the tragedy.[4]

Such stories and accounts of past events were part of the community memory in many areas of rural Ireland for many generations. They were passed on carefully by repeated telling and their significance thereby did not diminish over time. In some places story tellers used to get together and recite them to one another. But if they deviated in the detail they would be corrected and expected to adhere to the agreed, accepted version. Thereby, the telling of whirlwind events, from dust devils to tornadoes, became normalised as passing detail within these stories. This gives further confidence in establishing the reality of there being a long history of such vortices across Ireland.

William Butler Yeats found that, while the telling of these stories was open and unhindered within rural communities, this was not so with strangers.[5] What he experienced in his interactions with local people is still not unusual during site investigations today. He commented, 'You must go adroitly to work, and make friends ... for the fairies are most secretive, and much resent being talked of.' Such resentment by the faeries was considered as being the cause of many otherwise unexplained and sudden ailments and misfortunes among local people. The fear was a genuine one and was shared by successive generations. Today, this effect still lingers among some eyewitnesses of tornadoes, especially in the west and north-west of Ireland. This results in a reluctance to talk about what they have seen and experienced.

From popular story telling in all parts of the country, Irish people knew there was a risk to their own wellbeing if they encountered a *sidhe gaoithe*. They knew they were vulnerable, even though such events were few and far between in any one individual's experience. But they were always a possibility. If one did occur, people knew they should avoid it at all costs, since the consequences would be harmful and possibly fatal. This traditional sense of vulnerability provided both an awareness and a response mechanism that was considered necessary for living in rural environments. These include the lakes of Ireland where waterspouts occur from time to time, both as eddy whirlwinds and as tornadoes passing over water. Indeed, most of the lakes in Ireland were said to have a guardian *sidhe*, indicating a history of occasional waterspouts. These guardian *sidhe* were also known to have caused fatalities.

Whirlwinds of all kinds were dreaded because they carried crops and animals away. Some were said to have carried souls. Many were known to carry human beings away. In addition, they were also seen as the cause of severe hurt, injury and even deformity to ordinary people. If a whirlwind occurred suddenly, with all manner of debris being whirled high into the air, rural people knew to lie down on the ground for its duration. Anyone who did not and dared to stand in its path would know from what they had been taught that they could suffer various injuries and even facial deformity. For example, if you were caught in such a situation, it was thought possible that your mouth would become crooked and remain so for the rest of your life. Traditionally, in Ireland, the faeries were blamed for all such afflictions. Indeed, it was a traditional saying in some communities that one should never look in the direction of the whirlwind in order to avoid the *poc si* (faery stroke) which could result. In other communities the reaction to seeing a whirlwind and to the likelihood of being in its path was one of fear. This was expressed in terms which community members hoped would not offend and thereby lay them open to terrible retribution of some kind. So, they would take off their hats and say 'God bless them'! This apparent attitude of respect is reported widely in the literature and can only be understood when one is aware of the widespread sense of immediacy of the faery presence.

As with tornado experiences reported in our modern day, tornadoes in the past were sometimes heard before they were actually seen. It was very common for ancient descriptions of such events to mention a strange sound like the humming of a thousand bees before the whirlwind actually appeared.[6] This terrifying 'faery wind' was said to be a host of faeries passing by. They had the power to move quickly through the air and change shape quite dramatically.

The way in which these whirlwind events have been incorporated into core traditional folklore demonstrates how much a part of the Irish environment whirlwinds of all shapes and sizes have been for centuries, although not all of them would have been tornadoes or waterspouts as we know them today. The continuity of these accounts from generation to generation provided a mechanism for communities to learn about their vulnerability. It was not just a matter of identity and preserving a culture. As a result of this greater awareness and potential readiness to respond the vulnerability of people was actually lessened.

These traditions faded, but did not disappear, as Ireland was Christianised and colonialised. The new ways of thinking about the world tended to emphasise the past *sidhe* events as folklore tales and myth in a rather dismissive way. Nevertheless, an effort was made by some to record many of these events, but using a context that had a different world view. So, although whirlwinds of every size continued to occur, they were described in other terms by those who recorded them to pass on to later generations. As a result, the historical continuity of the record was gradually broken and references to these whirlwinds became increasingly set aside as myth or legend. Particularly as the population became more urbanised, and gradually detached from their rural cultural roots, successive generations began to learn about their world in different ways.

Vulnerabilities Expressed in the Early Irish Mirabilia

A new type of historical record of events in Ireland came with the arrival of Christian missionaries from other European regions and the subsequent establishment of churches in communities across the country. Soon afterwards, a further important influx came from colonial excursions from various parts of Europe that led to the introduction of very different ways of thinking about the world and ways of recording events.

Some of the historical records created by these communities include events that may be interpreted as tornadoes or, over water, waterspouts. These were written down and recorded as 'mirabilia', or wonders. It has been stressed that these were not claimed as 'miracula', or miracles, but 'mirabilia', which refers to an event that exceeds one's knowledge and expectations of the natural world.[7] So it was seen as a natural event, but quite inexplicable. 'What on earth is this? What is going on? I don't understand!' Such a response is not unfamiliar to some modern eyewitnesses in Ireland today. So the sense of wonder that is emphasised is certainly one of awe, but also of mental bewilderment. These mirabilia were recorded in the *Irish Annals*.

Desperation and horror caused by dragons and water serpents

As detailed in Chapter 3, one of the earliest images used to describe an event that was probably a tornado was that of a dragon. The earliest of these goes back to the Irish saint, St Abban, probably in the early AD 600s.[8] This was a near miss from a waterspout. He was in in a boat with other passengers and crew, offshore from County Wexford. It was a very close encounter since the sea was thrown into such a turmoil around the vessel that it was endangering it. This was a life-threatening encounter. With little or no scope to take effective avoiding action, the desperation of those aboard caused them to resort to prayer. Thus, unlike some other recorded dragon experiences, an immediate threat was perceived (although the perception was different from what the threat actually was) and the only protective response that was considered appropriate was initiated (prayer). The eyewitnesses expected the dragon's body to have been washed up along the shore, but they were disappointed. From this experience, they do not appear to have learnt anything with regard to waterspout hazards that they could pass on to others who might have a similar life-threatening encounter. Their frame of reference for understanding the experience was inappropriate for that.

10.1 Lough Fenagh – the site of a possible sixth-century waterspout.

An earlier record dating from the sixth century may well have been a similar waterspout event. It was particularly notable because of a significant number of fatalities.[9] This was on Lough Fenagh, a freshwater lake in County Leitrim. The *Book of Fenagh* records it as being caused by 'a horrible monster' (or 'water serpent', as it has been translated). So significant was this to the horrified community that not only was the event recorded (when any similar event without loss of life would not be), but thereafter, one of the names by which the lough was known was *Loch ni Pesti* (Lake of the Water Serpent). The fear of this happening again reputedly led to local people keeping off the water for many years. The extreme nature of the Fenagh event suggests it may well have been a significant waterspout (Figure 10.1).

By the eighth century, occasional reports of dragon events were made with little comment, so none of these experiences give any information about how people reacted to them, neither in the short nor the long term. Therefore, various dragon events were recorded but with no detail other than that they impressed as being huge.[10] The record gives no sense of vulnerability or fear. But they were perceived to be sufficiently exceptional to be recorded.

People and animals among the airborne debris

One of the most quoted forms of mirabilia with tornado-like characteristics was the 'ship in the sky'. The earliest description of this image gives expression not merely to a reaction of wonder, but also indicates severe consequences. The Clonmacnoise ship in the sky of AD 743 vividly describes a man falling to earth waving his arms as if swimming (Chapter 3). This would be a fairly accurate description of someone falling from a height, as would be the case if among the debris lifted into the air one or more people were swept up by the vortex and then dumped. It is hard to image that the consequences would be other than fatal. Unlike the modern day, the population of the Clonmacnoise area was large, as it became a major centre of learning of European-wide significance. Spending a lot of their lives in the open, people were vulnerable to severe weather conditions. In contrast, other ship-in-the-sky narratives are focused mostly on the vortex itself rather than on the people in its vicinity and the potential dangers to which they could have been exposed. Such an experience can happen to anyone in the immediate proximity of a tornado funnel. Even in recent times such a threat has only narrowly been averted in Ireland. In October 2002, a tornado that descended suddenly on the County Donegal village of Dunfanaghy almost sucked away one eyewitness. He had left his workshop to helplessly watch the advancing tornado funnel as it levelled a wall and headed towards his home. Only by anchoring himself to a very secure post was he able to prevent himself from being lifted as he experienced a strong upward suction.

The people who were eyewitnesses to the Clonmacnoise event appear to have been very close to the funnel of the tornado as it reached the ground. Their apparent closeness seems almost unreal, especially as the account evokes little sense of fear or apprehension such as is often conveyed in modern eyewitness accounts.

There were a number of very similar ship-in-the-sky events reported over the period AD 743–956 in different *Annals*. Where any details are given, they show that there were multiple eyewitnesses and the only public reaction reported was to attempt a response when a life-threatening situation arose. Otherwise they were matter-of-fact accounts in which convention was a strong influence in how they were reported. Clearly, these conventions had little concern for the reaction and fears of local people.

The lack of fatalities from towers of fire

The 'tower of fire' at Kilbeggan in AD 1054 was vividly described in the *Annals*, but no fire effects were noted. They are so lacking that the notion of a fire can be dismissed. It may well have been that the vortex was perceived as a column of smoke. By the time it reached Kilbeggan it would have been clearly apparent to the townsfolk that this was no fire. Indeed, none of the thatch and timber houses caught fire and no other fire damage was mentioned in the report. However, passing through a populated area around a monastery would mean there were bound to be casualties. But in the various accounts of this event the only ones specifically mentioned were dogs, cats and other small animals.

One report focused on a greyhound, which must have belonged to the monastery. This was reported as having been taken up, with fatal consequences when it was thrown back down to the ground. Had any of Kilbeggan's population been affected in a similar manner, it is more than likely that they would also have been featured in the report. They were certainly very vulnerable. But they were probably able to avoid coming to harm from seeing the tornado approach across the flat terrain and taking cover from what they may have thought was the approach of a dangerous fire. However, it is possible that these details may well have only applied to the monastery, which would have been the source of information for the annalist. Some damage to the town is likely to have occurred, but no mention is made of this.

Vulnerable turf cutters

Since the term 'whirlwind' was only introduced by the Vikings, it was only much later that records of destructive or catastrophic wind events used it. The open boglands across Ireland were one significant environment where people would be extremely vulnerable to these. The *Annals* report how in AD 1488 at least one turf cutter in the bog near Tuaim-Mona (or Tumona), north of Roscommon town, was killed by a whirlwind and others were injured. Further northwards along its track the same whirlwind killed four others.[11] Turf cutting was a major activity in many parts of Ireland over centuries and the boglands were open with no shelter. But despite the openness of the landscape where turf was cut, being able to move out of the way of an advancing vortex would be severely hindered by the deep long trenches, turf walls and sheer tiredness from working up to fourteen hours a day that turf cutters traditionally worked (Figure 10.2). So, while the landscape was open for the vortex, it was not so open for those in it to flee in any direction from the tornado's sudden appearance. The extreme vulnerability of the turf cutters is further emphasised in that many of the survivors did not escape other effects which were so striking as to be recorded as well. In the Tumona event, many of the survivors suffered from deformities, indicating that they were caught in some of the most severe internal winds of the vortex, where cold wind, very low humidity and intense

10.2 Traditional turf cutting that created a working environment of deep trenches in the bogs of the west of Ireland.

speed combined to break down the fat molecular layer of their skin, with a speed and intensity that reflected the severity of their exposure.

Threats from strange showers and objects falling from the sky

A rather different example of mirabilia that may have been associated with a tornado was the fall from the sky of quite unexpected objects that did not belong in the atmosphere in the first place. These have been recorded in the *Annals* from time to time as one or more showers of blood, honey, wheat or silver. Each of these events lacked any link with a specific source area from which the substances had been derived. But the noteworthiness of where they were deposited was considerable.

Along with other occurrences they were sometimes taken as portents for future events. Indeed, monastic centres where the *Annals* were originally compiled belonged to a medieval church in Ireland that had strong apocalyptic expectations. As a result, the descents of such substances were a major source of alarm for the local population. This contributed to an impulse for recording celestial and meteorological events which otherwise could well have been passed over. The record of events that may now be considered as possible tornadoes, or tornado-related, has benefited from this interest. On the other hand, similar arrivals of substances and objects with no apocalyptic associations may well have been left out of the record.

One of these, rain depositing blood on the ground, would appear to be a quite remarkable phenomenon. But there is a significant historical record of such events in many parts of Europe. The early application of modern science to this was in Germany in 1847, when Ehrenberg's analysis of samples showed minute dust particles of animal and vegetable forms. He dismissed suggestions that the Sahara might be the source of these and was of the view that instead they were 'drawn by violent winds out of such places as dried swamps and after being carried long distances at a great height [the material] descends in the rain'.[12] Tatlock has shown that on different occasions the apparent appearance of such blood may have been due to either fungi or dust lofted within the vortex. The earliest reference to this phenomenon in the *Annals* is for AD 690 in the *Annals of the Four Masters*.

These showers of blood were seen as a threat because they gave rise to an expectation of serious events to come, as well as being apocalyptic. This threat was a community-wide one and not one that afflicted one or two individuals. But, although a certain level of fear and alarm may have been generated by such portents at the time, none of the events that they could have been pointing to were identified in the *Annals*, even though they were compiled a considerable time afterwards when hindsight would have been able to function most effectively.

Exceptionally, a small number of entries in the *Annals* noting such events may well be romantic embellishments to make an impressive story at the time the original record was made. Such likely circumstance is possibly behind the record in the *Annals of Clonmacnoise* for AD 759 of showers of silver, honey and wheat. This was at a time of imminent famine and was described in the narrative as being a result of the king's prayers. But generally these records can be regarded as having been made with integrity, striving for the highest accuracy possible at the time.

These and similar Irish shower events were regarded as significant and noteworthy outside Ireland. They were not merely confined to the record contained in the various *Annals*, as important as that was for the record as a whole. For example, it was reported in the *Michigan Argus* of 31 March 1876, by Professor Smith of Louisville, that there had been a shower of dried frogspawn from ponds or swamps in 1675. Perhaps a more prestigious report in the *Philosophical Transactions of the Royal Society* for 1695, published in London, of showers of butter, was more widely noted.[13] The continuity of these showers after the period covered by the *Annals* has strengthened the legitimacy of the historical record for these events in Ireland. From time to time in more recent centuries there have been similar events, as of showers of nuts and berries in Dublin in 1867, showers of hay in Monkstown in 1879 and dozens of fish in Comber, County Down, in 1928. Each of these is very likely to have been caused by vortices over water or land. Generally they have been of little threat. The exception was the 1867 Dublin event which occurred 'in great quantities and with great force', so that the police had to take shelter from the onslaught!'[14]

When the mirabilia came to be recorded as part of the *Annals*, the monks did so without knowing whether they were of significance for the bulk of the population whom they served. Certainly the members of the monastic orders involved would have been aware of them as would the Gaelic aristocracy of the time. But ordinary folk would not have been particularly considered. Of course, there were rare occasions when they were directly involved in one of these events, but in most instances their vulnerability and reaction was not something that was considered essential nor appropriate for the record.

The mirabilia recorded throughout the *Annals* as a whole did not cover the events of the entire island. Inevitably, the record focuses on the area served by the monastic centres. This was limited and it fluctuated over time. Thus, the record is very far from being comprehensive. Nevertheless, there is no reason to consider some Irish regions to have been more vulnerable than others to those events that can be identified as tornadoes or waterspouts.

The Beginnings of Formal Reporting

Over the centuries there has been a major disconnect in communication and awareness between different parts of Ireland. This was not only a reflection of the relatively detached rural and urban worlds, but also the detachment between regions within Ireland. So there was limited awareness and little understanding of events that were relatively rare and mostly confined within any one area. Communication developed more quickly through various media, ranging from newspapers to academic journals. However, when these engaged with tornado events, they emphasised their distinctiveness and in their different ways generated a message that the event was quite exceptional, so the extent of Ireland's ongoing vulnerability received little attention. As a result, the vulnerability of the Irish population to further events continued, since there was no growth in awareness of possible recurrences and how to respond. Not knowing that a real threat exists increases the vulnerability of people to that threat. However, one positive from the expansion in forms of communication was that the tornadoes that were reported were placed on record in a form that could be recognised and used in the future.

The newspaper record

Beyond Dublin there was no regular newspaper in Ireland until after 1715.[15] Before that, such newspapers that were available were mostly English imports. In addition, the newspaper-reading public was mostly urban and were largely coffee house users where papers were readily available. So any awareness of tornado events developing as a result of reports from this source was inevitably limited and mostly confined to the very few large urban centres.

One of these newspapers was the *Belfast Newsletter*. This commenced in 1737. It was primarily aimed at the sizeable Protestant population and acquired an educated landowning and commercial readership in particular. Its spread into rural areas was aided on occasions when a local schoolmaster would read its contents aloud to small, assembled groups, to keep them informed. It had acquired a significant readership by 1775, when a destructive tornado tracked along the north-western shorelands of Belfast Lough. From the upper slopes of Divis Mountain near Belfast the tornado created a track for more than 28 km to Island Magee and then out to sea as a waterspout for an unspecified distance. The newspaper report highlighted how, besides causing much damage, the tornado caused terror among the numerous farm labourers harvesting in the fields along the track, engulfing some people walking near Windmill Hill and, having lofted them off the ground, tossed them into a ditch.

For such relatively infrequent events to grow an appropriate sense of vulnerability to further future events, a record needs to be available to successive generations. In the Carrickfergus case this became possible with its inclusion in the definitive record of the history of the Carrickfergus region by Samuel McSkimin in 1811, thirty-six years later.[16] This reference was relatively brief and lacked a lot of the available detail. Strikingly, it was influenced by the strongly held popular local views that the 1775 event was a result of what was described as a conflict between Irish and Scotch faeries. This was only changed in a much fuller use of the original information in a later, edited and expanded edition by his great-granddaughter, Elizabeth McCrum, in 1909. But these records were not enough to create any common sense of vulnerability, for thereafter reference to the event faded and the warning it provided was lost, even among the limited social group who would have been aware of it.

This 1775 tornado was referenced again in the *Belfast Newsletter* in 1834. This was on the occasion of another significant tornado that hit the area of Kilroot, just two miles north of Carrickfergus. This event also caused terror as two fleeing farm workers were thrown over a hedge while panic and fatalities occurred among farm animals. In addition, structural damage was inflicted upon buildings, including thatch roofing being swept away. The potential for building a sense of vulnerability was considerable since the report ended with a reference to 1775. The final sentence of the report read, 'On the 2nd September 1775, a similar tornado passed over in nearly the same direction, with nearly similar awful effect'. But it stopped short of any consideration as to what this might mean with regard to future events and what this might mean for the ongoing vulnerability of the population of the region.

Further tornadoes have occurred within this same area between Belfast and Island Magee up to the present, although each was separated by several decades. The most recent one was in 2007. This started at Carrickfergus and was reported this time in the *Belfast Telegraph*, as well as other media. None of these reports indicated an awareness of past events and concern for the ongoing vulnerability of the local population. It is noteworthy that the reports describe striking characteristics of what eyewitnesses saw that were almost identical to previous events. Some of these were noted before the tornado struck, including the description of the parent cloud, some of the extreme features of the storm, and a preceding whirring noise. Other common features mentioned might have enabled people to flee at the last minute had they been recognised, but these were only realised after the tornado had passed; features such as objects being twirled around in the air and people, animals and objects being lifted up and carried away.

Whether any of these could have alerted the local population had they known of them from earlier events is doubtful. It is likely that this would happen only if people had been exposed to a similar event within a certain time period. Indeed, perhaps the most striking feature of the *Belfast Telegraph* report in 2007 was the repetitive use of the term 'freak'. Four times in 400 words it is used to describe the event in the text. It has now become commonplace for Irish newspapers to include the term in headlines when reporting Irish tornadoes.

The scientific record

The first Irish tornado to be reported and detailed in a scientific journal was an event that occurred in October 1752 when a tornado tracked across parts of Termonamongan and Urney in County Tyrone. It damaged farm buildings and homes and caused some injuries but there were no fatalities. The impact details and track were put together by the rector of Urney parish with two colleagues from outside the area. Unusually, he went to these lengths because he wanted 'the more perfect knowledge of all particulars' and so together they 'viewed and measured' the damage marks and 'examined minutely' village eyewitnesses. This appears to be the first thorough site investigation carried out in Ireland. An agreed report was prepared and submitted to Lord Cadogan F.R.S.

at the Royal Society of London. As a result it was read at their meeting on 11 January 1753 and subsequently published in its *Proceedings*.[17] However, while this provided an excellent report for a growing scientific community, particularly outside Ireland, it did little for raising awareness of this hazard in Ireland.

By the end of the nineteenth century nearly 200 stations in Ireland were reporting their rainfall observations for publication in *Symons's Monthly Weather Magazine*. Some were also sending a wider range of meteorological observations, both instrumental and visual. This was not an Irish publication, but the journal became a significant outlet for meteorological enthusiasts of all backgrounds in Ireland, where they could report their observations and discuss weather phenomena. From time to time some of the tornado events in Ireland were included, but these reports failed to develop a wider awareness across the country, or even be included in the developing database for Ireland's climate profile in the journal. In only the first ten years of the journal (it was published from 1866) reports from Ireland that were either definite or possible tornadoes included events in 1867, 1869, 1870, 1871 and 1875.

These were relatively detailed accounts. As well as meteorological details and severe damage, most reported injuries and fatalities where they were known. But all of the events appear to have been largely regarded as relatively infrequent curiosities. The stated objective of the journal was not to raise awareness, nor to explore the vulnerability of people, but rather it had (and still has) the sole objective of the advancement of science. Much less frequently, other non-Irish journals were also outlets for tornado reports. These also had little or no intention of enhancing public awareness of their threat, although they did add to the list of events not referenced elsewhere, such as a tornado in 1872 reported in *Nature*.[18]

The drain of these reports away from Ireland shows that there was an urgent need for a forum within Ireland that could provide an alternative focus for them and would make it possible to build up a greater countrywide awareness. The RIA was established as a multidisciplinary body embracing both the sciences and humanities. Thus it had a wider scope than the Royal Society. It provided a forum for reporting all kinds of scientific observations from 1785 onwards. A number of tornadoes were reported in reasonable detail, but they were treated individually, and opportunities were not taken to get a broader picture of the threat they could pose.

The cluster of events in and around the 1850s, especially where urban environments and their populations were involved, was one such opportunity. It is clear from these reports that the vulnerability of the local populations was severe. In some cases deaths and injuries were reported as well as severe and remarkable damage to both property and the environment. Having experienced and seen the severity of what such storms could inflict would be reason enough to consider just how vulnerable Ireland was.

The tornado of 1850 that ploughed its way through the centre of Dublin was investigated by the president of the RIA himself, Humphrey Lloyd (see Chapter 4). In his published report he stated that it was 'so unusual (I may say unexampled) in these climates'. The report gave no indication of any awareness of other tornadoes in Ireland's past. However, he explained that it was 'produced by rapidly ascending currents of air caused by the heating of a limited portion of the earth's surface under the action of the sun's rays. In the temperate zones, accordingly, it is never produced in winter'. Such an authoritative statement indicated a marked seasonal vulnerability, i.e. they do not happen in winter months, but they could happen during summer months. This indirectly suggested there could be another.

There was a different emphasis in 1851 when the RIA requested a report from Dr Daniel Griffin of Limerick (see Chapters 4 and 9) about the destructive tornado that had hit Limerick in October of that year. As well as the physical destruction, Dr Griffin included much detail about the extremely variable vulnerable situations in which people had found themselves, showing that even in places that might be considered safe, it was not necessarily so. It was apparent that anyone who had become aware of any distinguishing characteristic – the sudden roar, a smoke-like column in the air, flying debris – had needed to respond and take shelter quickly, for they were all vulnerable (although it was not spelt out quite in such terms). However, no mention was made of other Irish tornadoes and no expectation of other future events in Ireland was raised.

Then came the most severe tornado yet. A mere twelve days after the RIA had received the Limerick report at its meeting of 12 January 1852, an even more intense tornado ploughed its way through the town of Nenagh, in County Tipperary, on 24 January (see Chapter 4). It was the most severe and intense tornado in the Irish written record. But it only reached three local/regional newspapers. The RIA did not include it in its scientific record and the possibility of informing the wider public that such an extreme event could occur in Ireland was missed.

More regionally orientated academic journals also contributed to the process of establishing an Irish record of such events. One of these was the *Cuvierian Journal*, published in Cork from 1835. In 1851, firmly within this period of significant tornado activity, two tornadoes occurred during the afternoon of 4 September on the north side of the city. They were widely seen from about one mile away. The record of this event could not have been possible without very observant and relatively informed eyewitnesses. The eyewitnesses were not able to describe any surface impacts, but they did observe detailed visual characteristics of the storm cell's developing shape and behaviour. All this led to a presentation to the society, consisting of 'a wide range of educated gentlemen', which was then recorded in their minutes and subsequently published. But the information went further. A report of the Society's meeting was published in the *Cork Examiner*, the regional newspaper. This brought it to the attention of a wider section of the educated community, although this would still have been a small proportion of the whole population. It is a remarkable thing that in the burst of journal reports about the significant tornado activity in major urban centres, no reference to the Cork event was made, thereby limiting the awareness of the event across the country as a whole. The reports for this cluster of tornado events during the mid-nineteenth century demonstrate the significant disconnect that existed between centres of scholarship in Ireland. Even target audiences/readerships never got a full picture of their intensity, frequency and other features. So the public's awareness did not grow and ultimately they remained as vulnerable as before. Also, their links with external centres of scholarship, rather than national ones, tended to dictate the direction of information flow. The result was a much weaker awareness of the tornado threat within Ireland among all Irish people than would otherwise have been the case.

Meteorological reports and climatic descriptions

From 1860, the beginnings of a formal, systematic gathering and transmission of weather data was initiated around the Irish coast by the naval authority in London to provide warnings of storms that could be passed to vessels at sea. But it was the British Meteorological Office that developed the perceived data-gathering needs for Ireland, both before Independence and for many years afterwards. The aim was less related to recording localised weather features than to gather large-scale, representative data that could be used to establish the location and movement of countrywide weather systems. This was continued with the establishment of the Irish Meteorological Service in 1936.

As daily weather reporting increased, the emphasis was on weather systems that were widespread and frequent. There was little or no reporting methodology for highly localised weather phenomena, whether tornadoes or otherwise. The beginnings of formal data-gathering systems in the nineteenth century and early twentieth century gradually established a climatic framework/profile for the country as a whole, as well as for individual regions. This profile excluded tornadoes and left everyone in the dark about their Irishness – tornadoes no longer happened in Ireland!

Among the consequences of this was the way in which possible tornadic conditions were not considered as factors in building design requirements, land use planning and human awareness building, particularly as these evolved to meet Irish needs during the twentieth century. This could extend (as in more recent years) to the recognition and provision for personal traumas that occur from the intense experiences of being close to a moving vortex and the terror experienced in being caught up in the wild chaos of the event. The climatic database that was developed is also significant in being used to build up a profile of the Irish climate presented as a fundamental part of the Irish educational experience of children, as well as others engaged in educational training at various levels.

CHAPTER ELEVEN

Suddenly Exposed

Always expect the unexpected.

Tornadoes arrive with a suddenness that is intensified many times over by being so unexpected. Sudden exposure to such an unexpected hazard can create a sense of helplessness and desperation. Being exposed to a very real possibility of serious physical harm brings a deep emotional crisis that may cause panic and strong irrational responses. Such reactions have occurred over the centuries when people in Ireland have encountered and suffered harmful effects from tornadoes. The most extreme of these effects have been death and injury. Awareness of such dangers has been greatest when the knowledge of such a possibility has been built into popular understandings and expectations of the environment in which people have lived. Experience is a powerful teacher if it is embraced by individuals and communities as an awareness-raising process. If this occurs, suitable adaptive behaviours and responses may develop to lessen the vulnerability when similar events occur in the future. But often the records that were made contributed little to popular awareness. Rather they tended to disappear into archives.

A New Awareness

In the twentieth century it became possible to research more deeply and to communicate much more widely. This has helped to transform the vulnerability of communities and individuals, but there is still a long way to go. Research over the past few decades has given new insights into the vulnerability of people. Lessons have been learnt from past events that can make response to future events more effective and much safer. But a lot of that depends upon the awareness of people to the risks and appropriateness of different responses.

Taking refuge

Much has been learnt in recent decades about appropriate refuges when a tornado strikes. Instinct often takes over in extreme circumstances and this can lead to responses that, in hindsight, were far from achieving the desired improved safety. Should one stay? Get better shelter? Flee? The desired outcome is to reduce one's vulnerability the most.

By far the majority of people exposed to the tornado threat have taken shelter in their home, as they do for storms of any kind. This is the traditional response because people's lives in Ireland have been largely rural and centred on the homestead. Today, the very familiarity of the home makes it feel the safest place, especially if it is nearby when the crisis occurs. In Ireland it is most unusual for a home to have a basement, which would be the safest place if the threat was fully realised. It is quite clear from post-storm investigations that many, particularly adults, keep an eye on what is going on outside and do not avoid proximity to windows, doors and external walls. All of these are vulnerable to being broken down and giving the storm an entrance into the house. However, others have rightly perceived the greater protection available in central areas, particularly small internal rooms and even under the stairs and in hallways. Multiple wall intersections in the centre provide greater resistance to the wind loads.

But curiosity and the unique wonder factor of the event have sometimes overridden the need to seek safety. It has not been unusual for people to suddenly find themselves in a highly vulnerable position when an external door was exposed to intense tornadic winds away from the core vortex. This has occurred in a number of events when the homeowner has been observing the advance of the vortex to the last possible moment, even as far as a short distance from their garden, and then trying to open the house door but finding it impossible to do so – such being the force of the wind against it! They had to flee elsewhere.

Away from the home a number of options have presented themselves to people out and about. One feature scattered all over Irish landscapes that has become more numerous over the last two centuries, and provides a readily accessible, relatively enclosed, small space, is the bridge. These are sturdy, short-span structures that carry (or pass over) railways, roads and even water. The smaller it is, the more effective the shelter it provides seems to be. But in the proximity of a tornado those sheltering under a bridge have found winds to be extreme. Winds have converged into the archway, accelerating and drawing in highly dangerous accelerating debris. Even the smallest debris item can be fatal in such circumstances. In Ireland this type of shelter has been used in tornado situations by cyclists, walkers, fishermen and vehicle drivers, fortunately without the most extreme consequences. But these people experienced a vulnerability they had not expected.[1] Studies of tornado injuries and deaths reveal that 25 per cent of serious injuries and 83 per cent of deaths of people outside resulted from becoming airborne. Most major injuries (94 per cent) were due to being struck by debris, the head being the most common site of injury in these cases.[2]

Out in the open, the only shelter often available is, ironically, probably the best to use in all circumstances – the ditch. It is below ground level (like a basement) and the severest winds pass over it together with the wind-blown debris. Across Ireland's peat landscapes, lacking field ditches, peat extraction has left numerous similar trenches. There have been occasions when these have successfully sheltered workers from tornadoes, although there have been fatalities as well. Being in a vehicle has also seemed to be reassuringly safe to some. Post-event enquiries reveal that they thought they would be able to either drive away if the wind became impossible or outrun the vortex in any case. They failed to appreciate how easily the violent, rotating wind could flip the vehicle. So, on occasions drivers and passengers have stayed inside a vehicle instead of stopping to shelter in a ditch. When caught by the inflow winds they have been terrified and realised their mistake, now desperately hoping they would not encounter the central vortex. Fortunately in the small number of Irish cases the car, whether stationary or mobile (both have occurred), has only been lifted and not turned over. A car is never a safe refuge.

Caravans, mobile homes and holiday homes have sprung up in recent decades, in coastal and inland areas. They often substitute as home when safety is sought in storms. The tornado has had a devastating effect on some of these, at times with their occupants inside. As described earlier, tornadoes have carried caravans and mobile homes up over perimeter fencing, even when chained to the ground. Holiday homes are often not very substantial structurally. In all cases they are not a safe refuge, but occupants tend to stay put rather than seek alternative shelter nearby, from ditches to built structures.

There are significant differences between public buildings in the safety they offer during a tornadic storm. In modern Ireland built structures have become increasingly large. Taking refuge in a nearby building is not unusual in a storm, and this will often be in a relatively large building. These range from schools to shopping centres, to industrial areas and transport hubs of various kinds. Often their principal buildings are associated with tall walls and long-span frames, which are very weak against strong wind pressures. These kinds of buildings include school halls, warehouses, department stores, metal industrial buildings and agricultural barns. Fortunately, few of these have been hit by a tornado in Ireland, although there have been more close near misses.

The sound of a tornado

This is probably the most underestimated effect of a tornado event. As a cause of severe trauma, it is probably second in importance only to personal injury. But unlike physical damage there are relatively few opportunities in post-event investigations to record how sound was experienced, compared with logging and measuring physical damage. When it came, it was unexpected and quite outside the range of past experiences of intense sound.

Yet the sound of a tornado appears to be at the heart of the tornado experience. In particular, night-time events are dominated by it. But even during daytime events it is a source of great anxiety, fear and often terror. Any researcher encounters this frequently when gathering personal accounts. In Ireland, as elsewhere, witnesses have used many analogies to compare familiar sounds with that of the tornado, but usually there is extreme difficulty in trying to use these. It is not merely the extraordinary volume of the sound that characterises it, but the nature of the noise itself.

For many who encountered a tornado, the sound was the first point of awareness that something extraordinary was happening. It was sudden, deafening and described as being unlike any other sound ever heard before. Most witness reactions to this depended on what they instinctively thought it sounded like. There have been frequent occasions when it was thought to be 'like a locomotive thundering towards us', 'a plane coming down', 'a jet airliner skimming my roof', 'a massive bomb exploding', 'as in the most terrifying horror movie ever'. But others have said it was a noise which was quite unlike anything heard before and they could not describe it. Even for those whose first awareness of a tornado has been visual, surprise has been expressed at the suddenness of the noise impact and the alarm it generated in addition to the other dramatic effects.

In modern reports of tornado events in Ireland sound has mostly been incidental in the detail that has been reported and recorded. However, its linkage to the occurrence of personal traumatic stress is probably underestimated. In Ireland it has been a helpful clue for night events when people had a near miss because they were in bed but were suddenly seemingly engulfed by this noise and then, later, saw the impact of the tornado outside their residences.

During daytime events such intense and terror-inducing noise has led to instinctive responses to rush away from the immediate location to get away from it. There have been occasions when this led to rushing into a more exposed and vulnerable situation than where people were originally sheltering from the parent storm. If the intense sound of a tornado causes such a reaction, the tornado is close. Flying debris has often been swept far outside the vortex itself and would certainly characterise the zone in which the roar of the tornado is heard. Thus, any move to a nearby location perceived to be less terrifying can result in becoming vulnerable to being hit by this debris. Such an action would be ill-advised unless other safety concerns made it imperative. This has been one of the reasons why a number of terrified witnesses in Irish events reported that they stayed put.

Afterwards, in the medium to longer term, a significant consequence for some has been a deep sense of vulnerability whenever winds are high or, in other cases, when storm clouds are building in the sky. A frequent comment made during site investigations concerns the rapid onset of a new sense of fear whenever such circumstances have occurred in the days and weeks that followed. As in the past, folk making these comments have indicated that they have rarely confided this reaction to others, even relatives and friends. But it has lived with them ever since their tornado experience.

Vulnerable People

Being at risk to an advancing tornado has a number of dimensions to it. Two in particular strongly condition people's ability to react in modern Ireland. The first is an actual awareness of the tornado's proximity (knowing where it is and how close it is). But a second is the ability to react. Those who are unable to react, or who are very limited in their ability to respond effectively, are at greatest risk.

Knowing the tornado is there, and is approaching, is fundamental. In Ireland there are no media warnings or sirens as in some countries. Instead, the clues are mostly visual. However, because tornadoes are very infrequent in any one area, even these clues may be difficult to believe. For many, only short-distance views would be possible because of the irregular terrain and high skyline limiting long-distance views. However, clues might be more than just the sight of the vortex. Both the behaviour of large items of debris in the air, or the intense sound effects, may trigger a realisation of what is about to impose itself. Often past experience has prepared people for serious wind events, especially in wind-prone areas such as the coast, so people may be familiar with the sound of storm winds. But these usually build gradually. The intense suddenness of the wind changes can indicate that it is not the usual kind of storm. In particular, some of those versed in the *sidhe gaoithe* have been able to recall the unique sound known to precede the faery blast and have alerted others close by. However, the occasions when this is known to have happened have been relatively few in recent years. But any of these clues for an imminent tornado have provided a very welcome, though small, amount of extra time to assess the risks and to respond.

Demographic change

It is not merely the size, intensity, track dimensions or other physical characteristics of a tornado that determine the level of its threat to people in its path. The demographic and socioeconomic characteristics of the local population also influences who is at risk; some are at greater risk than others. Such different vulnerabilities may be evident at different stages of the threat, either before, during or after the tornado strikes.[3]

One way this has been expressed in Ireland has been when an expanding population spreads into adjacent or distant areas which have a different tornado frequency. There have been cases where such incoming residents became well networked into their new community and were much less vulnerable because of the support they received, both in the immediate intensely reactive build-up and particularly in the post-event responses. In contrast, those who adopted a more socially isolated lifestyle tended to be much more vulnerable. Such geographical variation is shown in Chapter 6. But of course residential migration may also be to an area with a lower tornado frequency.

Relative poverty also plays a role in increasing vulnerability. Long-term residents with fewer resources can be slower to renew and upgrade deteriorating residences. Such residences have been more damaged than those consistently renovated, particularly with regard to roofing and window deterioration. In contrast, those building newer residents have, in some instances, simplified aspects of their construction in order to manage the cost. The same has occurred with small estate developments where pressures to keep it simple (and less expensive) have been strong. Many second homes in holiday areas have increasingly fallen into this category in recent years. Second home seekers are not normally considered to be affected by poverty. However, the limitation on financial resources has a similar effect in creating a relatively vulnerable situation.

Among the most vulnerable people are those with limited mobility. Health, age and even the need to care for others, may hamper movement to shelter during the immediate impact of the tornado and in effective responses afterwards. There is always a certain amount of desperate haste, or even panic, in such circumstances. Remarkably, witnesses who have been limited in these ways during past Irish tornado events have all been within their homes when conditions were most extreme. This was not because they had anticipated a tornado, but rather that the weather was so threatening. This reflects a traditional sensitivity to frequent weather deterioration that anticipates the need for an appropriate protective response. As a result, this group is less vulnerable than others lacking such traditional sensitivity.

Trauma

Traumatic experiences, which occur when a person becomes psychologically overwhelmed, are extraordinary. Trauma occurs because these experiences overwhelm any ordinary person's adaptations to life's circumstances. People in Ireland invariably experience trauma of some kind when they are confronted by a tornado because they have no expectation of such an experience in their wildest dreams. For most, nothing in their past experience has given them such an expectation. So, besides a physical footprint, a tornado often creates a significant non-physical one. This phenomenon was unidentified and otherwise largely overlooked in Ireland until the twentieth century.

For many of those who have witnessed, or have been impacted by, the power of a tornado in different parts of Ireland, stress was acute at first. Accounts from those affected have shown that in some cases this stress lasted for many days. They have told how they could not stop thinking about the event. Sleeping difficulties at night were often matched by intense daytime anxieties. All of this was new to them and in some cases affected them in their place of employment as well as at home. Indeed, they reported that other daylight activities were affected, such as sport, recreation and community activities.

On occasions, witnesses have had these symptoms return after they had initially died down and become dormant. On returning, much later, they were a form of PTSD, often triggered by another storm. Some have become anxious when reminded of the event in any way, such as by thunder rumbling, clouds developing or even trains passing. Friends have mentioned how a person has become introverted, isolated or preoccupied when minor background sounds have startled them (like a door closing). The fear of another storm is strong, and they imagine what may happen. Research in the USA has shown that all these effects can be mitigated, and resilience built up over time.

Vulnerable children

In Ireland, many tornadoes have impacted homes or outdoor events where children and adolescents have been present. Then, in the aftermath of a tornado, attention has been focused upon adults and their various responses to the tornado's impacts. In the majority of cases examined over the past two decades, children were provided with what comfort seemed necessary during the event itself, but afterwards they have been considered as requiring little further attention. It was generally the case that they had been provided for by taking them to some form of shelter, care given to them that would provide some reassurance appropriate for their age, and assurances that all will be well. As part of this parents have often avoided exposing children to the most severe consequences of the event. All these strategies are generally thought to be appropriate in most cases. But the aftereffects have sometimes been underestimated.

Children in many parts of Ireland where traditions have remained strong may be less vulnerable to full PTSD from tornado events. In many regions they have been subjected regularly to very severe winds from an early age. More particularly, many have learnt from an early age some of the traditional stories associated with whirlwinds of all sizes, from dust devils to the most destructive windstorm of 1839 that hit Ireland and for a long time has been a common reference point in the storm narratives within the country – most of them referencing the significance of the *sidhe gaoithe*.

Today, mature and elderly adults occasionally speak of past tornadoes which they experienced as children, or which their own parents or grandparents experienced, and passed that information on from one generation to another. Such accounts have emerged during many a field investigation in rural areas. These are deep memories that are undatable, despite attempts to do so. This, combined with the *sidhe gaoithe* traditions (e.g. never to speak about the *sidhe gaoithe* directly for fear of retribution), has resulted in events going unreported, particularly where the population is relatively dispersed.

As modern eyewitnesses, children have been a source of important information in a number of tornado events over the past twenty years. They have contributed not merely by verbal commentary, but also by sketching, reporting personal items being blown away, finding items out

11.1 Sketch of the Bailieborough tornadoes by a child eyewitness. Note the damaged rooftops (exposed beams) and felled trees. (Mark Gilmore)

of place in the countryside or timing information, as well as describing the behaviour of pets. For example, in one event a child's sketch revealed that there were two vortices, something that had escaped comment but was confirmed subsequently (Figure 11.1). In others, toys lost from house gardens were found at a distance across fields, helping to track the vortex. Also, children have found clothing items on nearby pathways enabling their original locations to be found. At times, children and teenagers are very positive co-investigators. This may well help them normalise their potentially traumatic experience, with a positive outcome.

However, it appears that many children and young adults are now losing their traditional awareness as media and expert commentary on these events is increasingly derived from external sources unfamiliar with traditional rural culture and understandings. As they do so, it is possible that children become increasingly vulnerable to the trauma of not knowing that these kinds of events can be expected from time to time, adding to the threat of being overwhelmed by the sheer terror of the encounter. But if the adult population can reconnect with the Irish past, there can be a move towards a more positive and appropriate way of providing a suitable commentary that will prepare children and young adults for the full range of weather extremes, such as tornadoes, that will inevitably occur in the country and which they may encounter themselves on rare occasions.

CHAPTER TWELVE

An International Context

Mol an óige agus tiocfaidh sí.
(Old Irish Proverb: 'Under the shelter
of each other, people survive'.)

We learn from the experience of others. This includes communities and states, as well as individuals. Across the world there is a long history of past experiences of extreme weather and provision of support, advice and resources which can lead to the survival of those at risk. This emphasises the importance today for those who have already endured tornado experiences to assist in preparing others. Thereby they help others in their turn to manage the impact of tornado events. It is not an easy accomplishment because it is often stressed that tornadoes are very infrequent and localised. Indeed, in many cases they are even beyond the living memory of current residents. For most people in Ireland this has been a serious impediment to becoming aware of the potential threat and being able to manage the event.

North America

North America has long dominated the thinking of Irish people when it comes to tornadoes. It is still evident today. For example, the Ballickmoyler tornado of December 2013 was described by a group of local people as 'a whirlwind, something you'd see in America on the television'. This is partly due to the strong historical and family links that Irish people have with the USA. In the 2017 US Census 10.1 per cent of the population recorded a full or partial ancestry from Ireland.[1] Irish people were among the early Europeans to settle in North America and would have been among those who witnessed tornadoes there.

Initially small numbers of Irish arrived and settled along areas of the east coast from the 1700s, but in much larger numbers after the 1840s.[2] This was well after the first tornado recorded by settling Europeans. The first English-language record was of one that hit Rehoboth, Massachusetts, in August 1671.[3] From the east coast European settlement soon spread westwards and the Irish were strongly represented among the first wave of settlers that crossed the Appalachians and gradually reached what was to become Texas, going ever deeper into the heart of tornado country. The Spanish had arrived in Texas much earlier from the south and in the 1700s there was a small Irish presence among them as well. One was Hugh O'Connor, born in Dublin in 1734, who became the Spanish governor of Texas for 1767–70. An early Irish presence in tornado-prone areas meant a growing awareness among the Irish emigrants of devastating tornadic storm events from early days, which would transfer back to Ireland.

It took many years to discover the real extent of tornadoes in the USA. Vast expanses that were sparsely populated meant so many tornadoes went unobserved, even well into the twentieth century. Observation was, and still is, the key method for recording tornadoes. This is so unlike other climatic phenomena such as temperature, rainfall, wind, etc. These are measured by fixed standard instruments which may not even require a constant human presence. Tornadoes were unpredictable and short-lived and vast expanses of terrain had few inhabitants to notice them. Nevertheless, many were seen, experienced and described, which led to an important early publication that proved to be an early authority on the tornado phenomenon, by John Finley in 1888.[4]

Today, the heart of the USA experience of tornadoes is normally regarded as being Tornado Alley. The use of this term only goes back to 1952 when it was used in the title of a research project to study severe weather in Texas, Oklahoma and adjacent states. This core area of tornado activity is now well known. But the lack of a precise definition to the term Tornado Alley emphasises how subjective it is. Maps showing its location vary in the spread they present, ranging from eight states

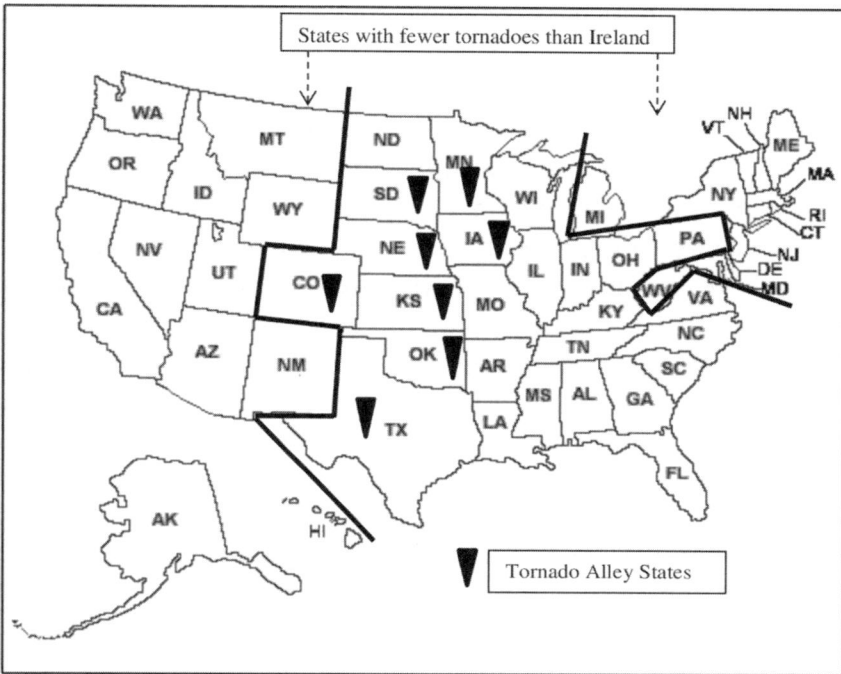

12.1 Ireland's theoretical location between Tornado Alley and those states with fewer tornadoes than Ireland.

(by the National Oceanic and Atmospheric Administration (NOAA)) up to eighteen states (storm chasers).[5] In one sense this matters little for this review because tornadoes occur in nearly every state and Irish people have long been aware of the Tornado Alley phenomenon as well as tornadoes in other parts of the USA. It was that awareness that was increasingly passed back to people in Ireland and shaped the thinking of many about the tornado phenomenon.

It provides a salient perspective on the Irish tornado experience to consider where it would fit into the USA experience. Based on the average number of tornadoes per year and allowing for size differences, Ireland's tornado frequency would exceed twenty-one of the states in the USA (Figure 12.1). This places Ireland geographically between the twenty-one states with lower frequencies and Tornado Alley (however defined). The twenty-one states fall into two regions. These are states in the western half of the country and those in its north-east. So, for example, in Figure 12.1, Ireland would fit between Wyoming and

Nebraska in the west and between Pennsylvania and New Jersey in the north-east.

Important contrasts between Ireland and the individual states that make up Tornado Alley go a long way towards explaining why there is such a large difference between them in terms of tornado numbers. The Great Plains area of the USA is vast. It has a terrain with relatively little variability. Unlike Ireland, there is a pattern of strong solar heating during the day and many much stronger frontal systems. As a result, the months in which tornadoes are most likely correspond to the times of year with increased solar heating and frequent strong frontal systems where cold air (from the north) and warm moist air (from the south) is channelled by mountain borders to the east and west of the region.[6] This is unlike Ireland where both warm and cold air has been significantly modified on its journey towards Ireland and then has to work its way over and through Ireland's rim of mountains and uplands. As a result, lower level convergence and uplift towards the jet stream above is much less dependent on temperature conditions. But tornadoes occur throughout the year in both the USA and Ireland.

Average totals only give a very partial picture of tornado activity. They vary considerably from year to year, both in the USA and Ireland. The start of the USA-wide tornado database was 1954, but even from then it is not as comprehensive as one might assume. Relatively weak tornadoes are significantly underrepresented in the early decades of the record. Before it was started many of the major events were documented when they occurred. This practice continued. But weaker tornadoes, even if they were seen, were relatively ignored because they did not have the impact of their larger and more impressive counterparts. However, it was experiences of the latter that were communicated onwards to people in Ireland and that shaped their awareness of what tornadoes were all about.

In modern times this deficit in observing and recording weaker tornadoes has largely been overcome. More comprehensive recording and reporting methods of the last few decades, including widespread public participation, have had a major effect on making the record more complete. It is widely agreed that as a consequence there has been an increase in tornado numbers during the last few decades. This has been

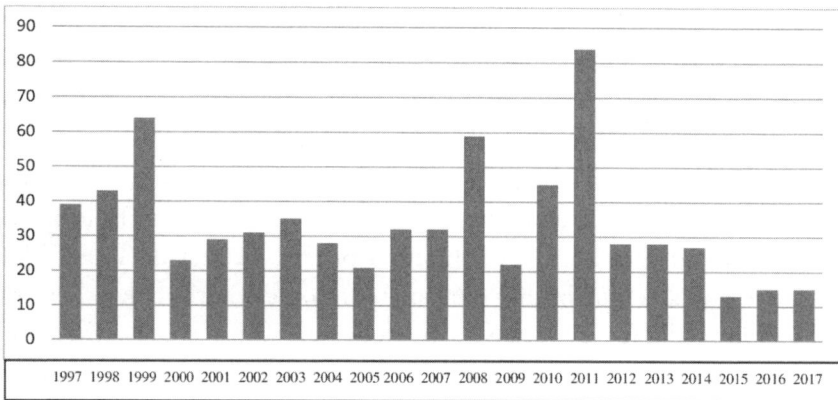

12.2 USA tornadoes 1997 to 2017 of F3 or greater.

mostly of weaker tornadoes, so the increase is more apparent than real.[7] This does not mean to say that every tornado gets recorded. As happens in Ireland, it is considered most probable that some are still missed.

This uncertainty with regard to the completeness of the record is not characteristic of stronger tornadoes in the USA. As a result, their annual totals over recent decades give a more assured indication of recent changes over time and any trending tendencies. For the same period for which the Irish data has been gathered (Chapter 7), the USA data for stronger tornadoes, F3 to F5, are shown in Figure 12.2. The lack of a trend is striking. Even if one considers the longer time period since the start of the record for these stronger tornadoes, the conclusion made by NOAA in 2014 was that 'there has been little trend in the frequency of the stronger tornadoes over the past 55 years'.[8] Perhaps more important than this is the high level of variability over such a short period. The data show that successive years can be very dissimilar, but not always. Only within very broad limits do they give any indication of what can be expected.

In some ways Ireland is a smaller mirror image of the USA experience for recent decades. Similar processes are at work, but their relative importance is not the same. Thus, such international comparisons help to highlight how the variety of processes that dominate the development of a tornado event may vary geographically and from country to country.

The European Context

Recording tornadoes in other European countries has had a very mixed history. As in Ireland, throughout the nineteenth century and the first half of the twentieth century there was active research in continental Europe. This produced a number of significant publications, particularly in German and French, as well as in the UK. On continental Europe it has been noted that after 1950 the level of this interest declined. Nor was there any attempt by national meteorological authorities there to build national databases because tornadoes were not deemed sufficiently significant. This was less so in the UK. Indeed, the first ever recognition of a supercell resulted from a study in the UK of the Wokingham tornado in 1962.[9] In 1974, the establishment of the private organisation TORRO provided a focus for a growing number of scattered individual weather observers seriously concerned to provide an accurate profile of UK and European tornadoes. In 2002, this stimulated the development of the ESSL (European Severe Storms Laboratory), which provided support for tornado researchers in mainland Europe, particularly with its associated ESWD (European Severe Weather Database).

But building databases and gaining a full understanding of tornadoes in Europe's environment are far from complete. Small steps towards this end have been taken. However, so far there has been little formal engagement with such research by many of the national meteorological services and other research institutions. Nevertheless the data that has been compiled from time to time in recent years has made possible early estimates of tornado frequencies across Europe as well as many records of individual significant events.

A recent European-wide data compilation has been done for the period 1950 to 2015.[10] This contains 5,478 confirmed tornadoes (including waterspouts that reached land) for the sixty-five-year period. At the national level Ireland would appear to have a relatively high frequency compared with other European nations (Figure 12.3). During this period Ireland registered 253 events. But in reality, the Irish total is likely to have been even greater than this. The pre-1997 data for Ireland is likely to have omissions as there was no comprehensive data gathering in the same way as occurred from 1997. This would place Ireland even

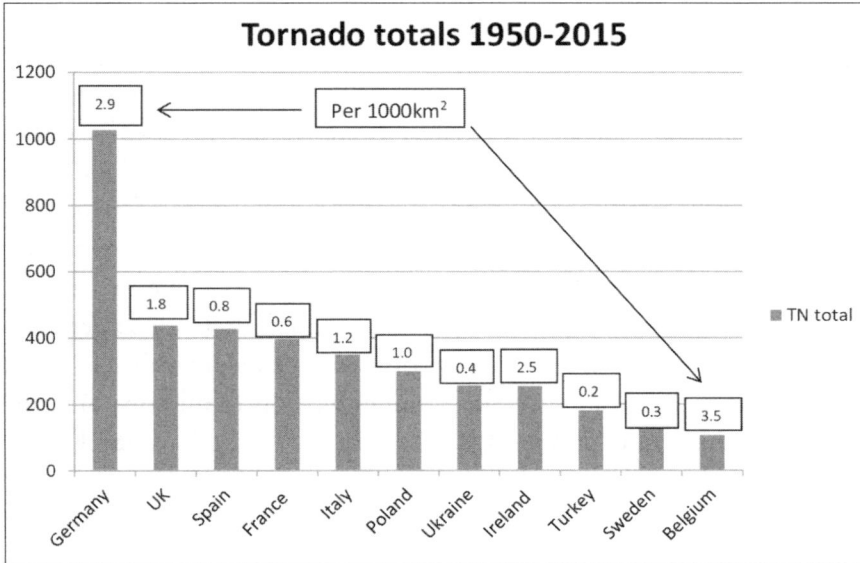

12.3 Tornado totals for European states, 1950 to 2015 and per 1000 km².

higher in the ranking. However, most countries would also have a significant shortfall in their data record for the same period.

Among the twelve states shown in Figure 12.3, Ireland ranks eighth by total number but has the third highest tornado numbers per 1000 km². To some extent this reflects its relatively small size. The countries with the highest totals are also much larger in area. Germany has the highest number of tornadoes (1027) for the period, but it is about four times the size of Ireland. So comparing them in terms of tornadoes per 1000 km², they are very similar. Germany has 2.9 and Ireland 2.5. One would expect a state with a large number of tornado events to have a greater level of public awareness. But both countries appear to have a significant public awareness deficit. Other pictures emerge from these data. Belgium has one of the highest tornado densities in Europe, at 3.5/1000 km². Compared with Ireland, it has had less than half the number of tornadoes. But it is only one third the size of Ireland. Despite these data, Belgium is never considered to be a country with a relatively high threat of tornadoes.

Despite updated estimates of annual tornado numbers for Europe by the ESSL, these data are still incomplete. There is widespread recognition of this within the European tornado research community. This incompleteness is significant and in some cases may well be up to 80 per cent. If this improved, the resulting picture would be much more realistic. In Ireland there is little public awareness of the wider European tornado frequency nor, as a result of that, of any potential threat in Europe. It is possible that if people in Ireland were aware of high levels of tornado threat among European nations, they would be more likely to engage with tornadoes as real possibilities in their own experience. Likewise, there is a lack in Europe's awareness of Ireland's experience. Many of the factors that have contributed to this are still having an effect, such as language diversity, communication limitations of many kinds and the results of historical interactions. At the local level, there is often no culture of reporting such events, especially if the events have had only limited impacts. Another major problem that emerges, even when there are reports, is the lack of an appropriate local verification process. The latter is critical for building a reliable database.

Since TORRO was formed in 1974, the UK is one of the few European states for which a fairly full coverage of tornado events has been recorded for recent years. A great deal of work has been done for earlier years as well. But the record for the last few decades has relatively few under-reporting problems, although they do exist. The twenty years from 1990 to 2010 are among the data published in a recent commemorative volume marking forty years of TORRO's existence.[11] Therefore, these are reasonably representative and cover part of the period for which annual totals are available for Ireland. Figure 12.4 shows that the period in question started with forty-seven tornadoes in 1990, after which the highest annual total was seventy-nine tornadoes in 2004 and the lowest was twenty tornadoes in 2008. The variations show no particular pattern, and this is likely to remain so with minor changes that may occur with the addition of any events in the under-reported areas. This is very similar to the data for the USA.

The current lack of a basic database for most European countries is being addressed by researchers in each individual country. As this progresses it may be possible to raise the awareness of the tornado hazard to national populations. The creation of a European-wide awareness then becomes a dimension of this because each state has a number of

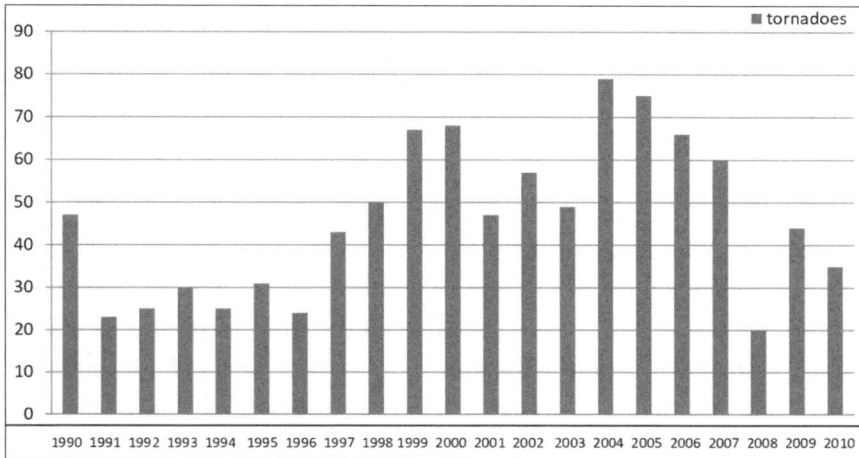

12.4 Annual tornado totals for the UK, 1990 to 2010.

neighbours and tornadoes do not respect borders. There is a need to know what a neighbour endures.

At this stage there is no compelling evidence of distinctive trends in tornado numbers. The data are not showing it and it has also been argued that even 'an increased frequency of favourable environments does not necessarily lead to an increased frequency of tornadoes resulting from those environments'.[12] Before such analyses can be attempted in any meaningful way there is an overarching need to understand much more fully the processes of tornado formation in the current and very diverse range of atmospheric conditions that occur in the different regions of Europe.

National exposure and awareness

A search for national tornado data sets reveals that a large proportion of them are, as yet, significantly incomplete. But those data that have been assembled are rather like the tip of an iceberg. What is seen and known indicates a reality that may be much greater. It is clear that nearly every state in the USA and every state within Europe, including the UK and Ireland, have tornadoes every year. In all cases the data that exist fluctuate on a year-by-year basis. Therefore, at the national and continental level there is a provisional minimum and maximum frequency that can be readily known. Even if tornadoes occur annually with a statistical randomness, certain conclusions can be drawn. It can be stated with certainty that tornadoes will occur in Ireland during the forthcoming year. Further it can be stated that there will be at least two. Beyond this it can also be stated that there may be up to fifteen. Exceptionally, this maximum total may be exceeded. But not knowing this higher figure is no reason not to recognise the inevitability of tornadoes and not to respond in an appropriate way.

At the beginning of this chapter it was stated that a major barrier to an effective response has been that tornadoes are infrequent and localised. However, this seems to be a localised perspective. At a wider level this is not so. For example, at a national level tornadoes are frequent, because they occur every year. What is not known is the when and where within that reality.

If tornadoes occurred with a random frequency there would be a noticeable irregularity in the annual event totals, because of some clustering over time. Likewise, if tornadoes occurred randomly across a state, they would not be evenly distributed. Some of them will cluster geographically. But where there are clusters in one year will not be where there are clusters in another year. This in itself creates a challenge for being prepared and expecting events to occur in this way. This is how the pattern of events in Ireland has occurred, in contrast to the recognition of something like Tornado Alley in the USA, where there is a continuing long history of frequent tornado development within the same broad region. But many countries are like Ireland and have events each year that are limited in size, duration, frequency and impact. Responses to these are a different challenge.

Giving a priority to knowing the future may result in a failure to learn from the past. Forward projections of chronologies will inevitably involve a high level of simplification of what are in reality relatively complex processes. Developing them is usually an expression of more than mere scientific curiosity. In the past this may not always have been so, but motivations have changed a great deal. Now they are more often linked to making provision to safeguard and protect, thereby minimising impacts. International comparisons show that the latter does not need to depend on whatever a forward projection might be. Knowing that an event will happen requires an adequate response system that will prepare people appropriately and will provide an effective response. There are lessons to be learnt from the past and from other communities and nations who are addressing similar challenges. There are also opportunities to make creative use of the accumulated knowledge that comes from a careful examination of each event.

Endnotes and References

CHAPTER 1 The Quest

1 All measurements are expressed in metric units, but often they have been converted from imperial units, as in this case.

CHAPTER 2 Encounters with Tornadoes

1 The words of Tom Grazulis, who as an eleven-year-old was awe-struck when the Worcester, Massachusetts tornado of 9 June 1953 passed close to his home. From his *Significant Tornadoes 1680–1991: A chronology and analysis of events* (St Johnsbury, VT: Environmental Films, 1993).

2 PTSD studies have ranged widely with regard to natural disaster events. General reviews of these have referenced tornadoes, e.g. Sandro Galea, Arijit Nandi and David Vlahov, 'The Epidemiology of Post-Traumatic Stress Disorder after Disasters', *Epidemiologic Reviews*, vol. 27, 2005, pp.78–91. One specific tornado study is Karen L. Midleton, Jonathan Willner and Kevin M. Simmons, 'Natural Disasters and Post Traumatic Stress Disorder Symptom Complex: Evidence from the Oklahoma tornado outbreak', *International Journal of Stress Management*, vol. 9(3), 2002, pp. 229–36. A number of phobia studies have been carried out, such as that by Margo C. Watt and Samantha Di Francescantonio, 'Who's Afraid of the Big Bad Wind? Origins of Severe Weather Phobia', *Journal of Psychopathology and Behavioral Assessment*, vol. 34, 2012, pp. 440–50.

3 Much data on the way the distinctive noise of the tornado precedes its arrival was published in David Hoadly, 'Tornado Sound Experiences', *Storm Track*, vol. 6(3), 1983. See also Alfred J. Bedard Jr., 'Low Frequency Atmospheric Acoustic Energy Associated with Vortices Produced by Thunderstorms', *Monthly Weather Review*, vol. 133(1), 2005, pp. 241–63.

4 These events are summarised in John G. Tyrrell, 'The Investigation of Tornadoes in Ireland, 2007', *The International Journal of Meteorology*, vol. 33(330), 2008, pp. 197–201.

5 From Helen Dukas and Banesh Hoffman, *About Einstein, the Human Side: New glimpses from his archives* (New York: Princeton University Press, 1979).

CHAPTER 3 The Early History of Irish Tornadoes

1 Jonathan Swift (1704), Angus Ross and David Woolley (eds), *A Tale of a Tub and Other Works*, Oxford World's Classics (Oxford: Oxford University Press, 1999), p. 16.

2 Daniel P. McCarthy, 'The Chronology of the Irish Annals', *Proceedings of the Royal Irish Academy. Section C: Archaeology, Celtic Studies, History, Linguistics, Literature*, 1998, pp. 203–55.

3 Alfred P. Smyth, 'The Earliest Irish Annals: Their first contemporary entries and the earliest centres of recording', *Proceedings of the Royal Irish Academy. Section C: Archaeology, Celtic Studies, History, Linguistics, Literature*, 1972, pp. 1–48.

4 Kuno Meyer, 'The Irish Mirabilia in the Norse "Speculum Regale"', *Folklore*, vol. 5(4), 1894, pp. 299–316.

5 John G. Tyrrell, 'Irish Tornadoes: Second thoughts about the Irish climate', *Journal of Meteorology, UK,* vol. 24, 1999b, pp. 404–11.

6 Careful analysis of the text shows that the dates in the *Annals of Ulster* for this period are out by one year and should be one year later than shown in the manuscript. The dates given here, and subsequently, allow for this adjustment. See Gearóid Mac Niocaill, *The Medieval Irish Annals* (Dublin: Dublin Historical Association, 1975). The two manuscripts are published as Conell Mageoghagan and Rev. Denis Murphy (eds), *The Annals of Clonmacnoise, Being Annals of Ireland from the Earliest Period to AD 140* (Burnham-on-Sea: Lanerch Publishers, 1993) and Cathal Maguire, William M. Hennessey and Bartholomew McCarthy (eds), *Annals of Ulster: A chronicle of Irish affairs, 431–1131 AD* (Dublin: HMSO, 1887).

7 Dorothy A. Bray, *A List of Motifs in the Lives of the Early Irish Saints* (Helsinki: Academia Scientiarum Fennica, 1992).

8 Donnchadh Ó Corráin, 'Mythology', in Brian De Breffny (ed.), *Ireland: A cultural encyclopaedia* (London: Thames & Hudson, 1983). Jakob Streit, *Sun and Cross: The development from megalithic culture to early Christianity in Ireland* (Edinburgh: Floris Books, 1984).

9 Jacqueline Simpson, *British Dragons* (London: Batsford, 1980).

10 Joseph Needham, *Science and Civilisation in Ancient China,* vol. 3, p. 479 (Cambridge: Cambridge University Press, 1959). For example he quotes a whirlwind that developed into a waterspout given by Yang Yü in *Shan Chii Hsin Hua* for 1360, 'On the fifteenth day of the twelfth month in the eighth year of the Chih-Chĕng reign period (+1348), at about three o'clock in the afternoon, there appeared in the south four black dragons coming down from the clouds and taking up water. Shortly afterwards another one appeared in the south-east and lasted a considerable time before it disappeared. This was seen at Chia-hsing city'. Mike Baillie, *Exodus to Arthur: Catastrophic encounters with comets* (London: Batsford, 2003). Baillie presents the accounts to support a comet theory and quotes (loosely) 'there were dragons fighting in a pool in Northern Liang Province ... trees were broken' (AD 503), 'Dragons fought in the pond of King of K'uh o ... they went westward as far as Kien ling ch'ing. In the places they passed all the trees were broken.' (AD 524). But they fit the tornado interpretation made by others.

11 David N. Dumville, *Histories and Pseudo-Histories of the Insular Middle Ages* (London: Routledge, 1990).

12 For a detailed consideration of alternative possible interpretations see Daniel P. McCarthy, *Chronological Synchronisation of the Irish Annals* 4th Edition (Trinity College, Dublin 2005). See www.cs.tcd.ie/Dan.McCarthy/chronology/synchronisms/annals-chron.htm

13 John O'Donovan (ed.), *Annals of the Kingdom of Ireland, by the Four Masters, from the Earliest Period to the Year 1616* (Dublin: Hodges & Smith, 1851).

14 Elliott B. Gose, *The World of the Irish Wonder Tale* (Toronto: University of Toronto Press, 1985).

15 Kenneth H. Jackson, *A Celtic Miscellany: Translations from the Celtic literatures* (London: Routledge and Kegan Paul, 1951).

16 Kathleen Hughes, 'Sanctity and Secularity in the Early Irish Church', in Derek Baker (ed.), *Sanctity and Secularity: The church and the world,* (Oxford: Blackwell, 1973, pp. 1–37). Also Bray (1992).

17 Kathleen Hughes, *The Early Celtic Idea of History and the Modern Historian* (Cambridge: Cambridge University Press, 1977).

18 Terence Meaden, 'Tornadoes in Britain: Their intensities and distribution in space and time', *Journal of Meteorology, UK,* vol.1, 1976, pp. 242–51.

19 Richard Best, Osborn I. Bergin and M.A. O'Brien (eds), *The Book of Leinster* (Dublin: Dublin Institute for Advanced Studies, 1954).

20 Dáibhí Ó Cróinín, *Early Medieval Ireland, 400–1200* (London, New York: Longmans, 1995).

21 Donnchadh Ó Meachair, *A Short History of County Meath* (Navan: An Uaimh, Meath Chronicle, 1928). Daniel A. Binchy, 'The Fair of Tailtiu and the Feast of Tara' *Eriu*, vol. 18, 1958, pp. 113–38.

22 James H. Todd and Algernon Herbert (eds), *Leabhar Breathnach Annso Sis: The Irish version of the Historia Britonum of Nennius* (Dublin: Irish Archaeological Society, 1848).

23 Aubrey Gwynn, *The Writings of Bishop Patrick 1074–1084* (Dublin: Dublin Institute for Advanced Studies, 1955).

24 Smyth (1972); Dumville (1990).

25 William M. Hennessy (ed.), *Chronicum Scotorum: A chronicle of Irish affairs from the earliest times to A.D. 1135* (London: Longmans, 1866). This is a translation of a transcription by MacFirbis (1585–1670) of earlier ancient documents.

26 Michael W. Rowe, 'The Earliest Documented Tornado in the British Isles: Rosdalla, County Westmeath, Eire, April 1054', *Journal of Meteorology, UK*, vol.14, 1989, pp. 86–90.

27 Aidan Breen and Daniel McCarthy,, 'A Re-Evaluation of the Eastern and Western Records of the Supernova of 1054', *Vistas in Astronomy*, vol.39(3), 1995, pp. 363–79.

28 Aubrey Gwynn, Richard Neville Hadcock and David Knowles, *Medieval Religious Houses of Ireland* (London: Longmans, 1970). Ann Donohoe (ed.), *Clonbroney with Ballinalee* (Ballinalee: Ballinalee Guild ICA, 1997).

29 Dumville (1990).

30 Mageoghagan and Murphy (1993).

31 Humphrey O'Sullivan, *The Diary of an Irish Countryman, 1827–1835*, Tomas de Bhaldraithe (tr.) (Cork: Mercier Press, 1997).

32 Timothy O'Neill Lane, *Lane's English–Irish Dictionary* (Dublin: Sealy, Bryers & Walker; London D. Nutt, 1904). Lambert McKenna, *English–Irish Phrase Dictionary* (Dublin: M.H. Gill & Son, 1922). Tomas de Bhaldraithe, *English–Irish Dictionary* (Dublin: Oifig An tSoláthair, 1959).

33 Daniel MacDonald's picture is reproduced in Peter Murray, *Whipping the Herring: Survival and celebration in nineteenth century Irish art* (Cork: Crawford Art Gallery and Gandon Editions, 2006).

34 Conchúr Ó Síocháin, *The Man From Cape Clear*, Riobárd P. Breatnach (tr.) (Cork: Mercier Press, 1984).

CHAPTER 4 A Prisoner of Consensus Science

1 Humphrey Lloyd, 'Notes on the Meteorology of Ireland, Deduced from the Observations made in the Year 1851, under the Direction of the Royal Irish Academy', *Proceedings of the Royal Irish Academy*, vol. 5, 1853, pp. 411–98.

2 John G. Tyrrell, 'The Cork Tornado of September 1851', *Journal of Meteorology, UK*, vol. 23(227), 1998, pp. 87–91.

3 John G. Tyrrell, 'Unexpected Meteorological Extremes: The Limerick tornado of 1851', *Irish Geography*, vol. 30(2), 1997, pp. 58–67.

4 Dr Daniel Griffin's letter to the Secretary of the Council of the Royal Irish Academy describing the event was sent on 26 November 1851 and read at their meeting on Monday, 12 January 1852. See *Proceedings of the Royal Irish Academy*, 1850–1853, pp. 225–30.

5 Humphrey Lloyd and William Hogan, 'On the Storm Which Visited Dublin on the 18th April 1850', *Proceedings of the Royal Irish Academy*, vol. 4, 1850, pp. 515–22.

6 *Dublin Evening Mail*, Friday, 19 April 1850.

7 Francis C. Bayard, 'English Climatology 1881–1890', *Quarterly Journal of the Royal Meteorological Society*, vol. 18(84), 1892, p. 213. New Series.

8   Lisa Shields, *The Irish Meteorological Service: The first fifty years* (Dublin: Stationery Office, 1987).

9   Richard E. Peterson, 'Tornadic Activity in Europe: The last half century', *Proceedings 12th Conference on Severe Local Storms*, San Antonio, Texas. American Meteorological Society, Boston, 1982, pp. 63–6.

10 The under-reporting of tornadoes in Britain, France and other countries is addressed in Chapter 12.

11 For example, the damage caused by the Ballylickmoyler tornado of December 2013 in County Laois included 'the entire orchard of about 50 trees… [that] were completely uprooted and lost' (personal communication with eyewitness). In total the damage was assessed as being in the order of 40,000 euro.

12 John G. Tyrrell, 'An Airborne Encounter with a Whirlwind Over County Kerry, Ireland' *Journal of Meteorology, UK,* vol. 23(227), 1999a, pp. 87–91.

13 The inaccurate media descriptions of tornadoes are illustrated by an event near Ballyhaunis, County Mayo, described in the *Western People* as a 'A miniature typhoon'.

14 This description was carried in the *Western People* on Tuesday, 6 September 2005.

15 The *Finn Valley Voice* of Wednesday, 5 April 1995 explained that 'such a wind is common' and that 'Shebeen whirlwinds come from the force of the fairies as they pass on their funeral lament on their way for the burial of their dead under the sea'. The report used no other term.

16 This definition is taken from the *Oxford English Dictionary.*

17 Thomas S. Khun, *The Structure of Scientific Revolutions* (Chicago: University of Chicago Press, 1970). An excellent example of this is how the accepted framework for understanding clouds changed dramatically when Luke Howard presented an alternative perception and framework. See Richard Hamblyn, *The Invention of Clouds* (London: Picador, 2001).

18 The spectrum of the numerous comments by eyewitnesses during a subsequent site investigation was very wide. They included 'It was very spectacular at the time both in sound and sight', 'Filled me with wonder', 'it savaged the trees', 'dark, thundery clouds with a most unusual formation in them', 'I looked out of the clubhouse window and observed the tornado about 200 yards away. Did not think much about it'!

19 This is one of the events to which reference is made in John G. Tyrrell, 'Tornadoes in Ireland During 2000', *Journal of Meteorology, UK,* vol. 26, 2001, pp. 183–4.

20 Additional information about the Ballysadare tornado is given in Chapter 5.

21 Meaden (1976).

22 The F-Scale was launched in Ted T. Fujita, 'F-Scale Classification of 1971 Tornadoes', *Satellite and Mesometeorology Research Project Paper* 100, Department of Geophysical Sciences, University of Chicago, 1972 (unpublished). The EF Scale proposal was made to the National Weather Service from the Wind Science and Engineering Center, Texas, and is at www.depts. ttu.edu/nwi/Pubs/FScale/EFScale.pdf. For a discussion of the different scales see G. Terence Meaden, S. Kochev, L. Kolendowicz, A. Kosa-Kiss, Izolda Marcinoniene, Michalis Sioutas, Heino Tooming and John Tyrrell, 'Comparing the Theoretical Versions of the Beaufort scale, the T-Scale and the Fujita Scale', *Atmospheric Research*, vol. 83(2–4), 2007, pp. 446–9.

23 John G. Tyrrell, 'The Investigation and Reporting of Tornadoes and Related Events in Ireland, 2009', *International Journal of Meteorology*, vol. 35(352), 2010, pp. 263–7.

24 John G. Tyrrell, 'Site Investigations of Tornado Events', in Robert K. Doe (ed.), *Extreme Weather* (Chichester: John Wiley & Sons Ltd, 2016), pp. 91–104.

25 From the last Sunday of March to the last Sunday of October both Irish Standard Time (IST) and British Summer Time (BST), which are the same, are observed in Ireland. These equate to Coordinated Universal Time + 1 hour (UTC +1). Both time standards are given to accommodate an internationally diverse readership.

26 This proved to be a tornado with the intensity of T 4.

27 The details of this can be found in John G. Tyrrell, 'A Trial Storm-Chase Across Munster and South Leinster, Ireland, to Test a Tornado Advisory', *Journal of Meteorology, UK*, vol. 28(280), 2003b, pp. 205–12.

28 Paul K. Knightley, 'Severe Weather Forecasts in the British Isles', *International Journal of Meteorology*, vol. 31(310), 2006, pp. 213–17.

29 John G. Tyrrell, 'Site Investigation Results of Tornado Reports in Ireland, 2006', *International Journal of Meteorology*, vol. 32(322), 2007, pp. 266–70.

CHAPTER 5 The Making of a Tornado

1 A quotation from the Bible: Job 36:29.

2 Bernoulli's principle relates wind speed to the associated pressure changes.

3 A full description of this event is published in the foresters' journal: John G. Tyrrell, 'A Tornado on the Galtees', *Coillte Contact*, vol. 16(3), 2004.

4 For further details of this case see Chapter 8 where site investigations are discussed.

5 Ted T. Fujita, 'Detailed Analysis of the Fargo Tornadoes of June 20, 1957', *US Weather Bureau Technical Report*, 5, 1959. Severe Local Storms Project, University of Chicago.

6 Roger M. Wakimoto and James W. Wilson, 'Non-supercell Tornadoes', *Monthly Weather Review*, vol. 117, 1989, pp. 1113–40.

7 The technical term describing this characteristic is 'helicity'.

8 CAPE is a measure of the extent to which a parcel of air, if forced upwards for any reason, would be warmer than the measured air temperature. So it indicates the amount of energy a parcel of air would have available for convection after having been lifted upwards. If CAPE is positive the air parcel would be buoyant and have a tendency to continue upwards until this was no longer the case.

9 Keith A. Browning, 'Airflow and Precipitation Trajectories within the Severe Local Storms that Travel to the Right of Winds', *Journal of Atmospheric Sciences*, vol. 21, 1964, pp. 634–9.

10 Ernest J. Fawbush and Robert C. Miller, 'The Types of Air Masses in which North American Tornadoes Form', *Bulletin of the American Meteorological Society*, 1954, pp. 154–65.

11 Fujita (1959).

12 For a discussion of the recognition and development of these downdrafts see Paul Markowski, 'Hook Echoes and Rear Flank Downdrafts: A review', *Monthly Weather Review*, vol. 130, 2002, pp. 852–76.

13 Also see Chapter 4 for the significance of the Ballysadare tornado.

14 Matthew R. Clark and Douglas J. Parker, 'Synoptic-scale and Mesoscale Controls for Tornadogenesis on Cold Fronts: A generalized measure of tornado risk and identification of synoptic types', *Quarterly Journal of the Royal Meteorological Society*, vol. 146(733), 2020, pp. 4195–225. Also see, Matthew R. Clark and David J. Smart, 'Supercell and Non-supercell Tornadoes in the United Kingdom and Ireland', in Robert K. Doe (ed.), *Extreme Weather: Forty years of the Tornado and Storm Research Organisation (TORRO)* (Chichester: John Wiley & Sons, 2016), pp. 31–59. David J. Smart and Keith A. Browning, 'Morphology and Evolution of Cold-frontal Miso-cyclones', *Quarterly Journal of the Royal Meteorological Society: A journal of the atmospheric sciences, applied meteorology and physical oceanography*, vol. 135, 2009, pp. 381–93.

15 Such as Matthew R. Clark, 'Doppler Radar Observations of Mesovortices within a Cool Season Tornadic Squall Line over the UK', *Atmospheric Research*, vol. 100(4), 2011, pp. 749–64.

CHAPTER 6 Waterspouts on Irish Waters

1   From the poem by Lucretius in his *De Rerum Natura* (On the Nature of Things) *Book VI*. The quotation is the beginning of a vivid description of waterspouts and how they are formed, indicating a familiarity that would have been common in the first century BC.

2   Charles Plummer, 'The Life of St Abban', in *Bethada Náem nÉrenn* (Oxford: Clarendon Press, 1922). Also see Edward Culleton, *Celtic and Early Christian Wexford: AD 400 to 1166* (Dublin: Four Courts Press, 1999).

3   Quoted in John F. Campbell, *The Celtic Dragon Myth* George Henderson (tr.) (California: Newcastle Publishing Co., 1981).

4   William R. Le Fanu, *Seventy Years of Irish Life* (London: Edward Arnold, 1893).

5   See Sharon Green's contribution to the Tornado Project at www.tornadoproject.com

6   The Gallery of Natural Phenomena at www.phenomena.org.uk/tornadoes/page6/page6.html

7   Humphrey O'Sullivan (1997).

8   I am grateful to Dr Kieran Hickey for these and other details of the Armagh record.

9   The details of this event are recorded more fully by Cormac Levis, *Towelsail Yawls: The lobsterboats of Heir Island and Roaringwater Bay* (Cork: Galley Head Press, 2002).

10 Estimates by different authorities for the length of Ireland's coastline (excluding offshore islands) vary between 2,797 km (1738 miles) to over 6,000 km (3,728 miles) due to different baselines used. As a result, estimates for its territorial waters range from 61,534 km² (23,758 square miles) to over 120,000 km²(46,332 square miles). Those quoted are the lowest values.

11 Russell Thompson, *Atmospheric Processes and Systems* (New York: Routledge, 1998).

12 Edward M. Brooks, 'Tornadoes and Related Phenomena', in Thomas F. Malone (ed.), *Compendium of Meteorology* (Boston, MA: American Meteorological Society, 1951).

13 Michalis Sioutas, Wade Szilagyi and Alexander Keul, 'Waterspout Outbreaks Over Areas of Europe and North America: Environment and predictability', *Atmospheric Research*, vol. 123, 2013, pp. 167–79. Ioannis Matsangouras, Panagiostis T. Nastos, Ioannis Pytharoulis and Mario M. Miglietta, COMECAP e-book of Proceedings, vol. 2, 2014, pp. 192–7.

14 Joanne Simpson, B.R. Morton, Michael C. McCumber and Richard S. Penc, 'Observations and Mechanisms of GATE Waterspouts', *Journal of Atmospheric Sciences*, vol. 43, 1986, pp. 753–82.

15 G.D. Hess and K.T. Spillane, 'Waterspouts in the Gulf of Carpentaria', *Australian Meteorological Magazine*, 1990, vol. 38, pp. 173–80.

16 Joseph H. Golden, 'An Assessment of Waterspout Frequencies Along the U.S. East and Gulf Coasts', *Journal of Applied Meteorology*, vol. 16, 1977, pp. 231–6; Joseph H. Golden and M.E. Sabones, 'Tornadic Waterspout Formation Near Interesting Boundaries', *Preprints, 25th International Conference on Radar Meteorology* (Paris, American Meteorological Society, 1991), pp. 420–3.

17 Grazulis (1993).

18 Simpson et al. (1986).

19 Joseph H. Golden, 'Some Statistical Aspects of Waterspout Formation', *Weatherwise*, vol. 26, 1973, pp. 108–17. Joseph H. Golden, 'The Life Cycle of the Florida Keys' Waterspouts', *Journal of Applied Meteorology and Climatology*, vol. 13(6), 1974a, pp. 676–92.

20 Joseph H. Golden, 'Scale-Interaction Implications for the Waterspout Life Cycle', *Journal of Applied Meteorology and Climatology,*, vol. 13(6), 1974b, pp. 693–709.

21 Reference to this event at Roundstone is also made in Chapter 5.

22 Howard Bluestein recalls that in the late 1970s as a tornado researcher of the Great Plains looking for effective monitoring devices, 'I was only vaguely aware of the earlier work on waterspouts'. See Howard B. Bluestein, *Tornado Alley: Monster Storms of the Great Plains* (New York: Oxford University Press, 1999), p. 94.

23 Golden (1974b).
24 Air Accident Investigation Unit Serious Incident: Europcopter SA 365N, EI-MIP, Kinsale Gas Field, 21 October 2004: Report No. 2006-008.
25 Two colour photos showing these features appear in the official report and may be viewed at http:www.aaiu.ie/sites/default/files/upload/general/12264-REPORT 2006_008-0.PDF.
26 Isaac Weld, *Illustrations of the Scenery of Killarney and the Surrounding Country* (London: Longman, Hurst, Rees, Orme & Carpenter, 1807).
27 The standard fishing boat on these lakes is a 17 ft (5.2 m) long timber boat. Information kindly supplied by Mike O'Donoghue and others.
28 John G. Tyrrell, 'Tornadoes, Other Whirlwinds and Thunderstorm Activity in Ireland, 2005' *International Journal of Meteorology*, vol. 31(310), 2006, pp. 191–4.

CHAPTER 7 Tornado Strikes in Ireland: Chronologies and patterns
1   The term 'Tornado Alley' was first used in 1952 by Ernest Fawbush and Robert Miller in their study of severe weather in parts of Oklahoma and Texas. It entered the public domain in the *New York Times* on 26 May 1957. The term is mostly used by the media to refer to a wide tornado-prone area between the Rocky Mountains and the Appalachian Mountains, but it has no clear definition.
2   For a summary of TORRO's history, with its stimulation of tornado research in Europe, see Terence Meaden, 'Researching Extreme Weather in the United Kingdom and Ireland: The history of the Tornado and Storm Research Organisation, 1974–2014', pp.1–13 in Doe (ed.) (2016).
3   The results of site investigations for many of these events were published in the *International Journal of Meteorology* (formerly known as the *Journal of Meteorology*).
4   Greg S. Forbes, 'Topographic Influences on Tornadoes in Pennsylvania', in *Preprints, 19th Conference on Severe Local Storms* (Minneapolis: American Meteorological Society, 1998), pp. 269–72.
5   For example, glider pilots are made aware of this hazard, as in C.E. Wallington, *Meteorology for Glider Pilots* (London: John Murray, 1961).
6   John G. Tyrrell, 'Site Investigation Data Applied to a Simple Tornado Developmental Model: the Carrigallen tornado of January 2002', *Journal of Meteorology, UK*, vol. 28, no. 275, 2003a, pp. 3–10.
7   This event is evaluated in John G. Tyrrell (2007).
8   The Nenagh tornado was recorded in detail in *The Nenagh Guardian* and the *Limerick Reporter*. It was also the subject of an unpublished undergraduate dissertation, Fergal Gleeson, *The Forgotten Tornado of Nenagh, January 24, 1852*, Department of Geography, University College Cork, 2008.
9   John G. Tyrrell, 'Site Investigation of a Multiple Tornado Event in Co. Westmeath, Ireland', *Journal. of Meteorology, UK*, vol. 27, no. 270, 2002, pp. 210–18.
10 John G. Tyrrell (2006), pp. 191–4.
11 Jonathan M. Davies and Anthony Fischer, 'Environmental Characteristics Associated with Nighttime Tornadoes', *Electronic Journal of Operational Meteorology*, 2009, pp. 1–29.
12 For example in Derek M. Elsom and Terence Meaden, 'Spatial and Temporal Distributions of Tornadoes in the United Kingdom, 1960–1982', *Weather*, vol. 39(10), 1984, pp. 317–23.
13 Edward Lorenz, 'Deterministic Nonperiodic Flow', *Journal of Atmospheric Sciences*, vol. 20, 1963, pp. 130–41.
14 Because of the discovery that weather prediction is extremely highly sensitive to a complex array of initial conditions the IPCC (Intergovernmental Panel on Climate Change) have themselves stated '... we should recognise that we are dealing with a coupled non-linear

chaotic system, and therefore that the long term prediction of future climate states is not possible'. See IPCC AR4 WG1 *Climate Change 2007: The physical science basis* (Cambridge and New York: Cambridge University Press, 2007).

CHAPTER 8 Investigating Irish Tornadoes

1 John G. Tyrrell (1999a), pp. 404–11. Further information about the Banemore tornado is in Chapter 4.
2 The meteorological conditions of this event are specifically discussed in Chapter 4, which also has a photograph of the tornado (Figure 4.7).
3 The St Patrick's Day tornado event is summarised in Irish Meteorological Service, 'Report on the Tornado in the Summerhill Area of County Meath, 17th March 1995' (Dublin: Irish Meteorological Service, 1995).
4 John G. Tyrrell (2002), pp. 210–18.
5 The deputy principal, Mr Michael Maguire, kindly coordinated the eyewitness evidence of students Kenneth Keegan, Richard Meares, Brendan Nannery, Brenda Watts and John McCormack. The detailed hourly weather diary observations made at Mount Russell, County Limerick by David Meskill and similarly by Martin Sweeney at Straide, County Mayo, were particularly helpful in researching this event.
6 Suction vortices are discussed in Chapter 5.
7 John G. Tyrrell, 'A Summer Outbreak of Whirlwind Phenomena from Dublin Bay to the Shannon Estuary', *Irish Geography,* vol. 37(1), 2004, pp. 20–36.
8 Joseph H. Golden and Daniel Purcell, 'Life Cycle of the Union City, Oklahoma Tornado and Comparison with Waterspouts', *Monthly Weather Review,* vol. 106(1), 1978, pp. 3–11.
9 John G. Tyrrell (2003a), pp. 3–10.
10 These are discussed in the UK context in Derek M. Elsom and Terence Meaden, 'Suppression and Dissipation of Weak Tornadoes in Metropolitan Areas: A case study of Greater London', *Monthly Weather Review,* vol. 110(7), 1982, pp. 745–56.
11 It is the convention to name a tornado by the location of its start point. Early site investigations for this event adopted this convention, but later work established that the tornado's track started further to the south-west. However, the original designation was retained to avoid confusion. Otherwise it has been referenced as the Ardmore tornado, Ardmore being the village closest to the start of the track.
12 Fred L. Haan, Jr., Partha P. Sarkar, Vasanth K. Balaramudu, 'Non-Stationary Tornado-Induced Wind Loads Compared with Traditional Boundary Layer Wind Loads on Low-Rise Buildings', *BBAA VI International Colloquium: Bluff bodies aerodynamics & applications, Milano, Italy,* 2008.
13 These were recognised initially in site investigations in the mid-1970s before being definitively reviewed in Ted Fujita, 'The Downburst, Microburst and Macroburst', *Satellite and Mesometeorology Research Project (SMRP)*, Research Paper 210, Department of Geophysical Sciences, University of Chicago (NTIS PB-148880), 1985.

CHAPTER 9 Trails of Destruction

1 This is described by Bernoulli's principle. See Horace Lamb, *Hydrodynamics* (Cambridge: Cambridge University Press, 1895).
2 Fred L. Haan, Vasanth K. Balaramuda and Partha P, Sakar, 'Tornado-Induced Wind Loads on a Low-Rise Building', *Journal of Structural Engineering*, vol. 136(1), 2010, pp. 106–16.
3 For the full details of this event see John G. Tyrrell, 'Unexpected Meteorological Extremes: The Limerick tornado of 1851', *Irish Geography*, vol. 30(2), 1997, pp. 38–67.

4 Department of the Environment and Local Government Statutory Instrument No. 497, 1997, Building Regulations 1997, indicates that the design wind speed determined in accordance with CP3: Chapter V: Part 2: 1972 as amended in 1986 should not exceed 44 m/s.

5 Haan et al. (2010).

6 The site investigation results for this event are in John G. Tyrrell, 'Local Effects and a Small Tornado in County Kildare, Ireland', *Journal of Meteorology, UK*, vol. 29(291), 2004, pp. 235–41.

7 See Chapter 4.

8 You-Lin Xu, *Wind Effects on Cable-Supported Bridges* (Singapore: John Wiley & Sons, 2013).

9 This effect was reported in the UK in Elsom and Meaden (1982), pp. 745–56.

10 The details of this event are given in Tyrrell (1997).

11 Stephen Cusak, 'Increased Tornado Hazard in Large Metropolitan Areas', *Atmospheric Research*, vol. 149, 2014, pp. 255–62.

12 A fuller account is given in Nicholas L. Betts, 'The Belfast Tornado', *Journal of Meteorology, UK*, vol. 8, 1983, pp. 78–80.

13 See Chapter 4 and Lloyd and Hogan (1850), pp. 515–22.

14 Anton Fischer, Philip Marshall and Ann Camp, 'Disturbances in Deciduous Temperate Forest Ecosystems of the Northern Hemisphere: Their effects on both recent and future forest development', *Biodiversity and Conservation*, vol. 22, 2013, pp. 1863–93.

15 See Chapter 1.

16 This distance is based on a relatively small amount of formal research, so longer distances are quite probable. See John A. Knox, Alan W. Black, John A. Rackley, Vittorio A. Gensini, Michael Butler, Minh Phan, Corey Dunn, Taylor Gallo, Melyssa R. Hunter, Lauren Lindsey, Robert Scroggs and Synne Brustad, 'Tornado Debris Characteristics and Trajectories during the 27 April 2011 Super Outbreak as Determined Using Social Media Data', *Bulletin of the American Meteorological Society*, vol. 94, 2013, pp. 1371–80.

17 This case is more fully reported in Don C. F. Cotton, 'A Fall of Sand Eels in Co. Sligo', *Irish Naturalists Journal*, vol. 27(10), 2004, pp. 407–8.

18 Some local views that they originated from a small lake about 4 km to the north of Dunlavin do not fit with the wind flow patterns nor with any of the eyewitness or site damage evidence of vortex activity.

19 O'Donovan (1851).

CHAPTER 10 Popular Responses Over Past Centuries

1 Lady Wilde, mother of Oscar Wilde, published her folklore records in Lady Jane Francesca Wilde, *Ancient Legends, Mystic Charms and Superstitions of Ireland* (London: Ward & Downey, 1888).

2 Oscar Wilde, *The Complete Fairy Tales* (Vermont: Norilana Books, 2007, 1st edition 1891).

3 For numerous such records see Simon Young and Ceri Houlbrook (eds), *Magical Folk: British and Irish fairies 500 AD to the present* (London: Gibson Square, 2017).

4 See Chapter 2. Also, Ó Síocháin (1984).

5 William B. Yeats, *Fairy and Folk Tales of Ireland* (Oxford: MacMillan Publishing Company, 1983). This is the combined publication of two of Yeats' works, *Fairy and Folk Tales of the Irish Peasantry*, 1888, and *Irish Fairy Tales*, 1892.

6 John O'Hanlon, *Irish Folklore: Traditions and superstitions of the country* (Glasgow: Cameron & Ferguson, Published under the pseudonym Lageniensis, 1870. Republished Salt Lake City: E.P. Publishing Ltd, 1973).

7 This distinction is made in Robert Bartlett, *The Natural and the Supernatural in the Middle Ages* (New York: Cambridge University Press, 2008) pp. 118–19.

8   Other details of this event are given in Chapter 6.
9   The exact year of this event is uncertain, but it was during the life of St Caillin, who recorded the event. See William M. Hennessey and D.H. Kelly (eds and tr.), *The Book of Fenagh in Irish and English* (Dublin: Alexander Thom, Abbey Street, 1875, reprint 1939).
10 William M. Hennessy (ed.), *Annals of Ulster, vol. 1*. (Dublin: Royal Irish Academy, 1887); D. Murphy (ed.), *The Annals of Clonmacnoise* (Dublin: Royal Society of Antiquaries of Ireland, 1896).
11 Dumville (1990).
12 Ehrenberg's work was summarised in English by Tatlock. See C.G. Ehrenberg (1847) *Passatstaub und Blutregen*, pp. 269–460, as quoted by John S.P. Tatlock, 'Some Medieval Cases of Blood Rain', *Classical Philology*, 9(4), 1914, pp. 442–7.
13 In *Philosophical Transactions of the Royal Society of London,* for 1695, there is an extract of a letter by Mr Robert Vans of Kilkenny, dated 15 November, reporting showers of butter-like or grease-like matter in the adjacent counties of Limerick and Tipperary: Robert Vans, and St George, 'An Account of an Extraordinary Meteor, or Kind of Dew Resembling Butter, That Fell Last Winter and Spring, in the Provinces of Munster and Leinster, in Ireland; Being Extracts of Two Letters, the One from Mr. Robert Vans to Mr. Henry Million, Dated November 15. 1695. The Other from the Right Reverend St George, Lord Bishop of Cloyne, to Sir Robert Southwell, U.P.R.S. Dated April 2. 1696. Wherein Mention Is Likewise Made of a Person Having a Regular Epileptick Fit Every Day at a Certain Hour.' *Philosophical Transactions of the Royal Society of London*, vol. 19, 1695, pp. 223–4.
14 The details are in a letter from Simon Pimm published in *Symons's Meteorological Magazine*, June, 1867, 2(13), p. 89.
15 The erratic beginnings of newspaper publishing in Ireland are traced by Robert Munton in *The History of the Irish Newspaper, 1685–1760* (New York: Cambridge University Press, 1967).
16 Samuel McSkimin, *The History and Antiquities of the County of the Town of Carrickfergus, from the Earliest Records to the Present Time* (Belfast: Hugh Kirk Gordon, 1811) p. 24. The later edition by his granddaughter was published in 1909, being *The History and Antiquities of the County of the Town of Carrickfergus, from the Earliest Records till 1839, by Samuel McSkimin*. New edition, with notes and appendix, by E.J. McCrum. (Belfast: Mullan & Son, James Cleeland, Davidson & M'Cormack).
17 William Henry, 'An Account of an Extraordinary Stream of Wind Which Shot Through Part of the Parishes of Termonomungan and Urney in the County of Tyrone, on Wednesday October 11, 1752. By Wm Henry, D.D. Rector of the Parish of Urney: Communicated by the Right Honourable the Lord Cadogan, F.R.S.' *Philosophical Transactions of the Royal Society of London*, vol. 48, 1753, pp. 1–4.
18 Originally a letter to the *Dublin Daily Express,* reproduced in *Nature*, 31 October 1872, p. 541.

CHAPTER 11 Suddenly Exposed
1   The acceleration of wind in this way is known as the Venturi effect, after Giovanni Venturi, 1746–1822.
2   Anne O. Carter, Margaret E. Millson and David E. Allen, 'Epidemiologic Study of Deaths and Injuries Due to Tornadoes', *American Journal of Epidemiology,* vol. 130(6), 1989, pp. 1209–18.
3   For the USA see William Donner and Havidán Rodriguez, 'Disaster Risk and Vulnerability: The role and impact of population and society' (*Population Reference Bureau: Washington DC, USA, 2011*).

CHAPTER 12 An International Context

1   US Census Bureau, '2013–2017 American Community Survey', www.census.gov/acs/www
2   Sallie A. Marston, 'Public Rituals and Community Power: St Patrick's Day parades in Lowell, Massachusetts, 1841–1874', *Political Geography Quarterly*, 8(3), 1989, pp. 255–69.
3   Grazulis (1993).
4   John P. Finley, *Tornadoes, What They Are and How to Escape Them* (Washington: J.H. Soule, 1888).
5   National Oceanic and Atmospheric Administration (NOAA) at www.ncdc.noaa.gov/climate-information/extreme-events/us-tornado-climatology
6   Bluestein (1999).
7   Kenneth F. Dewey, *Great Plains Weather* (Lincoln: University of Nebraska Press, 2019).
8   NOAA (2014).
9   Keith A. Browning and Frank Ludlam, 'Airflow in Convective Storms', *Quarterly Journal of the Royal Meteorological Society*, vol. 88, 1962, pp.117–35.
10   Bogdan Antonescu, David M. Schultz, Alois Holzer and Pieter Groenemeijer, 'Tornadoes in Europe: An underestimated threat', *Bulletin of the American Meteorological Society*, vol. 98(4), 2017, pp. 713–28.
11   Doe (ed.) (2016).
12   Patrick T. Marsh, Harold E. Brooks and David J. Karoly, 'Preliminary Investigations into the Severe Thunderstorm Environment of Europe Simulated by the Community Climate System Model 3', *Atmospheric Research*, vol. 93, 2009, pp. 607–18.

APPENDIX

Irish Tornadoes and Waterspouts Referenced in the Text

Year	Month	Location	County	Reference
500s	*	Lough Fenagh	Leitrim	Chapter 10
600s	*	Wexford coast	Wexford	Chapter 6; Chapter 10
734	*	Clonmacnoise area	Offaly	Chapter 3
735	*	Ulster	*	Chapter 3; Chapter 10
743	*	Clonmacnoise area	Offaly	Chapter 3; Chapter 10
740s	*	Clonmacnoise area	Offaly	Chapter 3
763	*	Teltown	Meath	Chapter 3
783	August	Clonbroney	Longford	Chapter 3
847	*	Slane	Meath	Chapter 3
950	*	Teltown	Meath	Chapter 3
984	*	Lough Hacket	Galway	Chapter 3
1054	April	Ross-deala	Westmeath	Chapter 3; Chapter 10
1488	*	Tumona	Roscommon	Chapter 10
1752	October	Termonamongan	Tyrone	Chapter 10
1775	September	Divis Mountain	Antrim	Chapter 10
1834	June	Kilroot	Antrim	Chapter 10
1850	April	Dublin	Dublin	Chapter 4; Chapter 9; Chapter 10
1851	September	Cork	Cork	Chapter 4; Chapter 10
1851	October	Limerick	Limerick	Chapter 4; Chapter 9; Chapter 10
1852	January	Nenagh	Tipperary	Chapter 4; Chapter 7; Chapter 10

(*not known)

APPENDIX *continued*

Year	Month	Location	County	Reference
1853	*	Kilkenny	Kilkenny	Chapter 10
1854	August	Ballinhassig	Cork	Chapter 10
1855	*	Quin	Clare	Chapter 10
1872	August	Lough Neagh	Antrim	Chapter 6
1886	*	*	Roscommon	Chapter 10
1894	April	Off South Coast		Chapter 3; Chapter 10
1908	December	Roaringwater Bay	Cork	Chapter 6
1978	September	Dungarvan	Waterford	Chapter 7
1982	September	Belfast	Antrim	Chapter 7; Chapter 9
1983	July	Roundstone	Galway	Chapter 5; Chapter 6
1983	July	Kippure Mountain	Wicklow	Chapter 9
1990	February	Foilmore	Kerry	Chapter 7
1995	February	Youghal	Cork	Chapter 1; Chapter 4; Chapter 7
1995	March	Castlefin	Donegal	Chapter 4
1995	March	Summerhill	Meath	Chapter 7; Chapter 8
1997	August	Cork city	Cork	Chapter 4
1998	January	Wexford town	Wexford	Chapter 4; Chapter 9
1998	June	Ballyclamper	Waterford	Chapter 9
1998	June	Dungarvan	Waterford	Chapter 7
1998	June	Stack's Mountains	Kerry	Chapter 4; Chapter 5
1999	July	Carraroe	Galway	Chapter 9
1999	August	Killeshandra	Cavan	Chapter 5
1999	August	Ballysadare	Sligo	Chapter 4; Chapter 5
1999	December	Coolrain	Laois	Chapter 7; Chapter 9
2000	June	Rosses Point	Sligo	Chapter 9
2000	July	Banemore	Kerry	Chapter 5; Chapter 8
2000	August	Clew Bay	Mayo	Chapter 6
2000	November	Brow Head	Cork	Chapter 7; Chapter 9
2000	November	Rossmore	Cork	Chapter 7
2000	November	Kilbeggan	Westmeath	Chapter 7
2000	November	Donaghadee	Down	Chapter 7
2001	April	Belmullet	Mayo	Chapter 8

APPENDIX *continued*

Year	Month	Location	County	Reference
2001	August	Dublin Bay	Dublin	Chapter 8
2001	September	Horseleap (2)	Westmeath	Chapter 5; Chapter 8
2001	September	Rosemount	Westmeath	Chapter 8
2001	September	Mullingar	Westmeath	Chapter 5; Chapter 8; Chapter 9
2002	January	Carrigallen	Leitrim	Chapter 7; Chapter 8
2002	October	Dunfanaghy	Donegal	Chapter 3; Chapter 10
2002	October	Preban	Wicklow	Chapter 4
2002	October	Castlederg	Tyrone	Chapter 4
2002	October	Derrynacross	Longford	Chapter 4; Chapter 9
2002	October	Derry Water	Wicklow	Chapter 9
2003	May	Rathangan	Kildare	Chapter 9
2004	January	Athlone	Westmeath	Chapter 9
2004	July	Galty Mountains	Tipperary	Chapter 5
2004	October	Bravo Platform	Cork	Chapter 6
2005	January	Killucan	Westmeath	Chapter 7
2005	January	Ballymore	Westmeath	Chapter 7
2005	January	Ballinamullen	Westmeath	Chapter 7
2005	January	Clonmore	Offaly	Chapter 7
2005	January	Clonee	Meath	Chapter 7
2005	January	Markethill	Armagh	Chapter 7; Chapter 9
2005	August	Inver	Mayo	Chapter 9
2006	March	Bailieborough	Cavan	Chapter 4; Chapter 11
2006	April	Togher	Meath	Chapter 8
2006	September	Bealistown	Wexford	Chapter 9
2006	December	Tolans Point	Armagh	Chapter 7; Chapter 8
2007	May	Carrickfergus	Antrim	Chapter 10
2007	July	Kinrush	Tyrone	Chapter 2
2008	February	Kinnegad	Westmeath	Chapter 5
2010	September	Robertstown	Meath	Chapter 5
2011	June	Straboe	Laois	Chapter 9
2011	December	Castleisland	Kerry	Chapter 9
2012	October	Crumlin	Dublin	Chapter 9

APPENDIX *continued*

Year	Month	Location	County	Reference
2013	October	Meelick	Galway	Chapter 8
2013	October	Clonfert	Galway	Chapter 8
2013	October	Lough Ree	Roscommon	Chapter 8
2013	December	Kilmallock	Limerick	Chapter 9
2013	December	Ballickmoyler	Cavan	Chapter 12
2014	June	Lough Neagh	Antrim	Chapter 6

Bibliography

Air Accident Investigation Unit. Serious Incident: Europcopter SA 365N, EI-MIP, Kinsale Gas Field, 21 October 2004: Report No. 2006-008

Baillie, Mike, *Exodus to Arthur: Catastrophic encounters with comets* (London: Batsford, 2003)

Bartlett, Robert, *The Natural and the Supernatural in the Middle Ages* (New York: Cambridge University Press, 2008)

Bayard, Francis, C., 'English Climatology 1881–1890', *Quarterly Journal of the Royal Meteorological Society*, vol. 18(84), 1892, New Series

Bedard, Alfred J. Jr., 'Low Frequency Atmospheric Acoustic Energy Associated with Vortices Produced by Thunderstorms', *Monthly Weather Review*, vol. 133(1), 2005

Best, Richard, Osborn I. Bergin and M.A. O'Brien (eds), *The Book of Leinster* (Dublin Institute for Advanced Studies, 1954)

Betts, Nicholas L., 'The Belfast Tornado', *Journal of Meteorology, UK*, vol. 8, 1983

Binchy, Daniel A., 'The Fair of Tailtiu and the Feast of Tara', *Eriu*, vol. 18, 1958

Bluestein, Howard B., *Tornado Alley: Monster Storms of the Great Plains* (New York: Oxford University Press, 1999)

Bogdan Antonescu, David M. Schultz, Alois Holzer and Pieter Groenemeijer, 'Tornadoes in Europe: An underestimated threat', *Bulletin of the American Meteorological Society*, vol.98(4), 2017

Bray, Dorothy A., *A List of Motifs in the Lives of the Early Irish Saints* (Helsinki: AcademiaScientiarum Fennica, 1992)

Brooks, Edward M., 'Tornadoes and Related Phenomena', in Thomas F. Malone (ed.), *Compendium of Meteorology* (Boston, MA: American Meteorological Society, 1951)

Breen, Aidan and Daniel, I. McCarthy, 'A Re-Evaluation of the Eastern and Western Records of the Supernova of 1054', *Vistas in Astronomy*, vol. 39(3), 1995

Browning, Keith A., 'Airflow and Precipitation Trajectories Within the Severe Local Storm that Travel to the Right of Winds', *Journal of Atmospheric Sciences*, vol. 21, 1964

Browning, Keith A. and Frank Ludlam, 'Airflow in Convective Storms', *Quarterly Journal of the Royal Meteorological Society*, vol. 88, 1962

Campbell, John F., *The Celtic Dragon Myth* George Henderson (tr.) (California: Newcastle Publishing Co., 1981)

Carter, Anne O., Margaret E. Millson and David E. Allen, 'Epidemiologic Study of Deaths and Injuries Due to Tornadoes', *American Journal of Epidemiology*, vol. 130(6), 1989

Clark, Matthew R., 'Doppler Radar Observations of Mesovortices Within a Cool Season Tornadic Squall Line over the UK', *Atmospheric Research*, vol.100(4), 2011

Clark, Matthew R. and David J. Smart, 'Supercell and Non-supercell Tornadoes in the United Kingdom and Ireland', in Robert K. Doe (ed.), *Extreme Weather: Forty years of the Tornado and Storm Research Organisation (TORRO)* (Chichester: John Wiley & Sons, 2016)

Clark, Matthew R. and Douglas J. Smart, 'Synoptic-scale and Mesoscale Controls for Tornadogenesis on Cold Fronts: A generalised measure of tornado risk and identification of synoptic types', *Quarterly Journal of the Royal Meteorological Society*, vol. 146(733), 2020

Cotton, Don C.F., 'A Fall of Sand Eels in Co. Sligo', *Irish Naturalists Journal*, vol. 27(10), 2004

Culleton, Edward, *Celtic and Early Christian Wexford: AD 400 to 1166* (Dublin: Four Courts Press, 1999)

Cusak, Stephen, 'Increased Tornado Hazard in Large Metropolitan Areas', *Atmospheric Research*, vol. 149, 2014

Davies, Jonathan M. and Anthony Fischer, 'Environmental Characteristics Associated with Nighttime Tornadoes', *Electronic Journal of Operational Meteorology*, 2009

de Bhaldraithe, Tomas, *English–Irish Dictionary* (Dublin: Oifig An tSoláthair, 1959)

Department of the Environment and Local Government, Statutory Instrument No. 497, 1997, Building Regulations, 1997

Dewey, Kenneth F., *Great Plains Weather* (Lincoln: University of Nebraska Press, 2019)

Doe, Robert K. (ed.), *Extreme Weather: Forty years of the Tornado and Storm Research Organisation (TORRO)* (Chichester: John Wiley & Sons, 2016)

Donner, William and Havidán Rodriguez, 'Disaster Risk and Vulnerability: The role and impact of population and society' (*Population Reference Bureau: Washington DC, USA, 2011*)

Donohoe, Ann (ed.), *Clonbroney with Ballinalee* (Ballinalee: Ballinalee Guild ICA, 1997)

Dukas, Helen and Banesh Hoffman, *About Einstein, the Human Side: New Glimpses from his Archives* (New York: Princeton University Press, 1979)

Dumville, David N., *Histories and Pseudo-Histories of the Insular Middle Ages* (London: Routledge, 1990)

Elsom, Derek M. and Terence Meaden, 'Suppression and Dissipation of Weak Tornadoes in Metropolitan Areas: A case study of Greater London', *Monthly Weather Review*, vol. 110(7), 1982

Elsom, Derek M. and Terence Meaden, 'Spatial and Temporal Distribution of Tornadoes in the United Kingdom, 1960–1982', *Weather*, vol. 39(10), 1984

Fawbush, Ernest J. and Robert C. Miller, 'The Types of Air Masses in Which North American Tornadoes Form', *Bulletin of the American Meteorological Society*, 1954

Finley, John P., *Tornadoes, What They Are and How to Escape Them* (Washington: J.H. Soule, 1888)

Fischer, A., P. Marshall and A. Camp, 'Disturbances in Deciduous Temperate ForestEcosystems of the Northern Hemisphere: Their effects on both recent and future forest development', *Biodiversity and Conservation*, vol. 22, 2013

Forbes, Greg S., 'Topographic Influences on Tornadoes in Pennsylvania', in *Preprints, 19th Conference on Severe Local Storms* (Minneapolis, MN: American Meteorological Society, 1998)

Fujita, Ted T., 'Detailed Analysis of the Fargo Tornadoes of June 20, 1957', *US Weather Bureau Technical Report*, 5, 1959. Severe Local Storms Project, University of Chicago

Fujita, Ted T., 'F-Scale Classification of 1971 Tornadoes', *Satellite and Mesometeorology Research Project Paper* 100, Department of Geophysical Sciences, University of Chicago, 1972 (unpublished)

Fujita, Ted, 'The Downburst, Microburst and Macroburst', *Satellite and Mesometeorology Research Project (SMRP)*, Research Paper 210, Department of Geophysical Sciences, University of Chicago (NTIS PB-148880), 1985

Galea, S., A. Nandi and D. Vlahov, 'The Epidemiology of Post-Traumatic Stress Disorder after Disasters', *Epidemiologic Reviews*, vol. 27, 2005

Gleeson, Fergal, *The Forgotten Tornado of Nenagh, January 24, 1852* Unpublished undergraduate dissertation, Department of Geography, University College Cork, 2008

Golden, Joseph H., 'Some Statistical Aspects of Waterspout Formation', *Weatherwise*, vol. 26, 1973

Golden, Joseph H., 'The Life Cycle of the Florida Keys' Waterspouts', *Journal of Applied Meteorology and Climatology*, vol. 13, 1974a

Golden, Joseph H., 'Scale-Interaction Implications for the Waterspout Life Cycle', *Journal of Applied Meteorology and Climatology*, vol. 13, 1974b

Golden, Joseph H., 'An Assessment of Waterspout Frequencies Along the U.S. East and Gulf Coasts', *Journal of Applied Meteorology and Climatology*, vol. 16, 1977

Golden, Joseph H. and Daniel Purcell, 'Life Cycle of the Union City, Oklahoma Tornado and Comparison with Waterspouts', *Monthly Weather Review*, vol. 106(1), 1978

Golden, Joseph H. and M.E. Sabones, 'Tornadic Waterspout Formation Near Interesting Boundaries', *Preprints, 25th International Conference on Radar Meteorology* (Paris: American Meteorological Society, 1991)

Gose, Elliott B., *The World of the Irish Wonder Tale* (Toronto: University of Toronto Press, 1985)

Grazulis, Thomas P., *Significant Tornadoes 1680–1991: A chronology and analysis of events* (St Johnsbury, VT: Environmental Films, 1993)

Grazulis, Thomas P., *The Tornado: Nature's ultimate windstorm* (Norman: University of Oklahoma Press, 2001)

Groenemeijer, Pieter and Thilo Kuhne, 'A Climatology of Tornadoes in Europe: Results from the European Severe Weather Database', *Monthly Weather Review*, vol. 42(12), 2014

Gwynn, Aubrey, *The Writings of Bishop Patrick 1074–1084* (Dublin: Dublin Institute for Advanced Studies, 1955)

Gwynn, Aubrey, Richard Neville Hadcock and David Knowles, *Medieval Religious Houses of Ireland* (London: Longmans, 1970)

Haan, Fred L. Jr., Partha P. Sarkar and Vasanth K. Balaramudu, 'Non-Stationary Tornado-Induced Wind Loads Compared with Traditional Boundary Layer Wind Loads on Low-Rise Buildings', *BBAA VI International Colloquium: Bluff Bodies Aerodynamics & Applications, Milano, Italy*, 2008

Haan, Fred L., Vasanth K. Balaramuda and Partha P. Sakar, 'Tornado-Induced Wind Loads on a Low-Rise Building', *Journal of Structural Engineering*, vol. 136(1), 2010

Hamblyn, Richard, *The Invention of Clouds* (London: Picador, 2001)

Hennessey, William M. (ed.), *Chronicum Scotorum: A chronicle of Irish affairs from the earliest times to A.D. 1135* (London: Longmans, 1866)

Hennessy, William M. (ed.), *Annals of Ulster, vol. 1* (Dublin, Royal Irish Academy, 1887)

Hennessey, William M. and D.H. Kelly (eds and tr.), *The Book of Fenagh in Irish and English* (Dublin: Alexander Thom, Abbey Street, 1875, reprint 1939)

Henry, William, 'An Account of an Extraordinary Stream of Wind Which Shot Through part of the Parishes of Termonomungan and Urney in the County of Tyrone, on Wednesday October 11, 1752. Communicated by the right honourable the Lord Cadogan, F.R.S.', *Philosophical Transactions of the Royal Society of London*, vol. 48, 1753

Hess G.D. and K.T. Spillane, 'Waterspouts in the Gulf of Carpentaria', *Australian Meteorological Magazine*, 1990, vol. 38

Hoadly, David, 'Tornado Sound Experiences', *Storm Track*, vol. 6(3), 1983

Houser, Jana L., Howard B. Bluestein and Jeffrey C. Snyder, 'A finescale radar examination of the tornadic debris signature and weak-echo reflectivity band associated with a large violent tornado', *Mon. Weather Rev.*, vol. 144(11), 2016

Hughes, Kathleen, 'Sanctity and Secularity in the Early Irish Church', in D. Baker (ed.), *Sanctity and Secularity: The church and the world* (Oxford: Blackwell, 1973)

Hughes, Kathleen, *The Early Celtic Idea of History and the Modern Historian* (Cambridge: Cambridge University Press, 1977)

Intergovernmental Panel on Climate Change, AR4 WG1 *Climate Change 2007: The physical science basis* (Cambridge and New York: Cambridge University Press, 2007)

Irish Meteorological Service, 'Report on the Tornado in the Summerhill Area of County Meath, 17th March 1995' (Dublin: Irish Meteorological Service, 1995)

Jackson, Kenneth H., *A Celtic Miscellany: Translations from the Celtic literatures* (London: Routledge and Kegan Paul, 1951)

Khun, Thomas, S., *The Structure of Scientific Revolutions* (Chicago: University of Chicago Press, 1970)

Knightley, Paul K., 'Severe Weather Forecasts in the British Isles', *International Journal of Meteorology*, vol. 31(310), 2006

Knox, John A., Alan W. Black, John A. Rackley, Vittorio A. Gensini, Michael Butler, Minh Phan, Corey Dunn, Taylor Gallo, Melyssa R. Hunter, Lauren Lindsey, Robert Scroggs and Synne Brustad, 'Tornado Debris Characteristics and Trajectories during the 27 April 2011 Super Outbreak as Determined Using Social Media Data', *Bulletin of the American Meteorological Society*, vol. 94, 2013

Lane, Timothy O., *Lane's English–Irish Dictionary* (Dublin: Sealy, Bryers & Walker; London D. Nutt, 1904)

Lamb, Horace, *Hydrodynamics* (Cambridge: Cambridge University Press, 1895)

Le Fanu, William R., *Seventy Years of Irish Life* (London: Edward Arnold, 1893)

Levis, Cormac, *Towelsail Yawls: The lobster boats of Heir Island and Roaringwater Bay* (Cork: Galley Head Press, 2002)

Lloyd, Humphrey, 'Notes on the Meteorology of Ireland, Deduced from the Observations made in the Year 1851, under the Direction of the Royal Irish Academy', *Proceedings of the Royal Irish Academy*, vol. 5, 1853

Lloyd, Humphrey and William Hogan, 'On the Storm Which Visited Dublin on the 18th April 1850', *Proceedings of the Royal Irish Academy*, vol. 4, 1850

Lorenz, Edward, 'Deterministic Nonperiodic Flow', *Journal of Atmospheric Sciences*, vol. 20, 1963

Lucretius, Titus, *De Rerum Natura* (On the Nature of Things) *Book VI*

Mac Niocaill, Gearóid, *The Medieval Irish Annals* (Dublin: Dublin Historical Association, 1975)

Mageoghagan Conell and Rev. Denis Murphy (eds), *The Annals of Clonmacnoise, Being Annals of Ireland from the Earliest Period to AD 140* (Burnham-on-Sea: Lanerch Publishers,1993)

Maguire Cathal, Hennessey and B. McCarthy (eds), *Annals of Ulster: A chronicle of Irish Affairs, 431–1131 AD* (Dublin: HMSO, 1887)

Markowski, Paul, 'Hook Echoes and Rear Flank Downdrafts: A review', *Monthly Weather Review*, vol. 130, 2002

Marsh, Patrick T., Harold Brooks and David J. Karoly, 'Preliminary Investigations into the Severe Thunderstorm Environment of Europe Simulated by the Community Climate System Model 3', *Atmospheric Research,* vol. 93, 2009

Marston, Sallie A., 'Public Rituals and Community Power: St Patrick's Day parades in Lowell, Massachusetts, 1841–1874', *Political Geography Quarterly*, vol. 8(3), 1989

Matsangouras, Ioannis, Panagiostis T Nastos, Ioannis Pytharoulis and Mario M. Miglietta, COMECAP e-book of Proceedings, vol. 2, 2014

McCarthy, Daniel P., 'The Chronology of the Irish Annals', *Proceedings of the Royal Irish Academy. Section C: Archaeology, Celtic Studies, History, Linguistics, Literature*, 98C, 1998

McCarthy, Daniel P., *Chronological Synchronisation of the Irish Annals* 4th Edition (Dublin: Trinity College, Dublin 2005)

McCrum, Elizabeth, *The History and Antiquities of the County of the Town of Carrickfergus, from the Earliest Records till 1839, by Samuel McSkimin*. New edition, with notes and appendix (Belfast: Davidson & M'Cormack, 1909)

McKenna, Lambert, *English–Irish Phrase Dictionary* (Dublin: M.H. Gill & Son, 1922)

McSkimin, Samuel, *The History and Antiquities of the County of the Town of Carrickfergus, from the Earliest Records to the Present Time* (Belfast: Hugh Kirk Gordon, 1811)

Meaden, Terence, 'Tornadoes in Britain: Their intensities and distribution in space and time', *Journal of Meteorology, UK*, vol. 1, 1976

Meaden, Terence, 'Researching Extreme Weather in the United Kingdom and Ireland: the history of the Tornado and Storm Research Organisation, 1974–2014', in Robert K. Doe (ed.), *Extreme Weather: Forty years of the Tornado and Storm Research Organisation (TORRO)* (Chichester: John Wiley & Sons Ltd, 2016)

Meaden, Terence, S. Kochev, Leszek Kolendowicz, Attila Kosa-Kiss, Izolda Marcinoniene, Michalis Sioutas, Heino Tooming and John Tyrrell, 'Comparing the Theoretical Versions of the Beaufort Scale, the T-Scale and the Fujita Scale', *Atmospheric Research*, vol. 83(2–4), 2005

Meyer Kuno, 'The Irish Mirabilia in the Norse "Speculum Regale"', *Folklore*, vol. 5(4), 1894

Midleton, Karen L., Jonathan Willner and Kevin M. Simmons, 'Natural Disasters and Post Traumatic Stress Disorder Symptom Complex: Evidence from the Oklahoma tornado outbreak', *International Journal of Stress Management*, vol. 9(3), 2002

Munton, Robert, *The History of the Irish Newspaper, 1685–1760* (New York: Cambridge University Press, 1967)

Murphy, D. (ed.), *The Annals of Clonmacnoise* (Dublin: Royal Society of Antiquaries of Ireland, 1896)

Murray, P., *Whipping the Herring: Survival and celebration in nineteenth century Irish art* (Cork: Crawford Art Gallery and Gandon Editions, 2006)

National Oceanic and Atmospheric Administration (NOAA): www.ncdc.noaa. gov/climate-information/extreme-events/us-tornado-climatology

Needham, Joseph, *Science and Civilisation in Ancient China*, vol. 3 (Cambridge: Cambridge University Press, 1959)

Ó Corráin, Donnachadh, 'Mythology', in Brian De Breffny (ed.), *Ireland, A cultural encyclopaedia* (London: Thames & Hudson, 1983)

Ó Cróinín, Dáibhí, *Early Medieval Ireland, 400–1200* (London, New York: Longmans, 1995)

O'Donovan, John (ed.), *Annals of the Kingdom of Ireland, by the Four Masters, from the Earliest Period to the Year 1616* (Dublin: Hodges and Smith, 1851)

O'Hanlon, John, *Irish Folklore: Traditions and superstitions of the country* (Glasgow: Cameron & Ferguson, Published under the pseudonym Lageniensis, 1870. Republished Salt Lake City: E.P. Publishing Ltd, 1973)

Ó Meachair, Donnchadh, *A Short History of County Meath* (Navan: An Uaimh, Meath Chronicle, 1928)

Ó Síocháin, Conchúr, *The Man From Cape Clear*, Riobárd P. Breatneach (tr.) (Cork: Mercier Press, 1984)

O'Sullivan, Humphrey, *The Diary of an Irish Countryman, 1827–1835*, Tomas de Bhaldraithe (tr.) (Cork: Mercier Press, 1997)

Peterson, Richard, E., 'Tornadic Activity in Europe: The last half century', *Proceedings 12th Conference on Severe Local Storms*, San Antonio, Texas. American Meteorological Society, Boston, 1982

Plummer, Charles, 'The Life of St Abban', in *Bethada Náem nÉrenn* (Oxford: Clarendon Press, 1922)

Rowe, Michael W., 'The Earliest Documented Tornado in the British Isles: Rosdalla, County Westmeath, Eire, April 1054', *Journal of Meteorology, UK*, vol. 14, 1989

Shakespeare, William, *Antony and Cleopatra, The Cambridge Dover Wilson Shakespeare* (John Dover Wilson (ed.), Cambridge: Cambridge University Press, 2009)

Shields, Lisa, *The Irish Meteorological Service: The first fifty years* (Dublin: Stationery Office, 1987)

Simpson, Jacqueline, *British Dragons* (London: Batsford, 1980)

Simpson, Joanne, B.R. Morton, Michael C. McCumber and Richard S. Penc, 'Observations and Mechanisms of GATE Waterspouts', *Journal of Atmospheric Sciences*, vol. 43, 1986

Sioutas, Michalis, Wade Szilagyi and Alexander Keul, 'Waterspout Outbreaks Over Areas of Europe and North America: Environment and predictability', *Atmospheric Research*, vol. 123, 2013

Smart, David J. and Keith A. Browning, 'Morphology and Evolution of Cold-frontal Miso-cyclones', *Quarterly Journal of the Royal Meteorological Society: A journal of the atmospheric sciences, applied meteorology and physical oceanography*, vol. 135, 2009

Smyth, Alfred P., 'The Earliest Irish Annals: Their first contemporary entries and the earliest centres of recording', *Proceedings of the Royal Irish Academy. Section C: Archaeology, Celtic Studies, History, Linguistics, Literature*, 1972

Streit Jakob, *Sun and Cross: The development from megalithic culture to early Christianity in Ireland* (Edinburgh: Floris Books, 1984)

Swift, Jonathan (1704), Angus Ross (ed.) and David Woolley (ed.), *A Tale of a Tub and Other Works* Oxford World's Classics (Oxford University Press, 1999)

Tatlock, John S.P., 'Some Medieval Cases of Blood Rain', *Classical Philology*, 9(4), 1914

Thompson, Russell, *Atmospheric Processes and Systems* (New York: Routledge, 1998)

Todd, James, H. and Algernon Herbert (eds), *Leabhar Breathnach Annso Sis: The Irish version of the Historia Britonum of Nennius* (Dublin: Irish Archaeological Society, 1848)

Tyrrell, John G., 'Unexpected Meteorological Extremes: The Limerick tornado of 1851', *Irish Geography,* vol. 30(2), 1997

Tyrrell, John G., 'The Cork Tornado of September 1851', *Journal of Meteorology, UK*, vol. 23(227), 1998

Tyrrell, John G., 'An Airborne Encounter with a Whirlwind Over County Kerry, Ireland', *Journal of Meteorology, UK*, vol. 23(227), 1999a

Tyrrell, John G., 'Irish Tornadoes: Second thoughts about the Irish climate', *Journal of Meteorology, UK*, vol. 24(228), 1999b

Tyrrell, John G., 'Tornadoes in Ireland During 2000', *Journal of Meteorology, UK*, vol. 26, 2001

Tyrrell, John G., 'Site Investigation of a Multiple Tornado Event in Co. Westmeath, Ireland', *Journal of Meteorology, UK,* vol. 27(270), 2002

Tyrrell, John G., 'Site Investigation Data Applied to a Simple Tornado Developmental Model: The Carrigallen tornado of January 2002', *Journal of Meteorology, UK*, vol. 28(275), 2003a

Tyrrell, John G., 'A Trial Storm-Chase Across Munster and South Leinster, Ireland, to Test a Tornado Advisory', *Journal of Meteorology, UK,* vol. 28(280), 2003b

Tyrrell, John, 'A Tornado on the Galtees', *Coillte Contact*, vol. 16(3), 2004

Tyrrell, John G., 'Local Effects and a Small Tornado in County Kildare, Ireland', *Journal of Meteorology, UK*, vol. 29(291), 2004

Tyrrell, John G., 'A Summer Outbreak of Whirlwind Phenomena from Dublin Bay to the Shannon Estuary', *Irish Geography*, vol. 37(1), 2004

Tyrrell, John G., 'Tornadoes, Other Whirlwinds and Thunderstorm Activity in Ireland, 2005', *International Journal of Meteorology*, vol. 31(310), 2006

Tyrrell, John G., 'Site Investigation Results of Tornado Reports in Ireland, 2006', *International Journal of Meteorology*, vol. 32(322), 2007

Tyrrell, John G., 'The Investigation of Tornadoes in Ireland, 2007', *International Journal of Meteorology*, vol. 33(330), 2008

Tyrrell, John G., 'The Investigation and Reporting of Tornadoes and Related Events in Ireland, 2009', *International Journal of Meteorology*, vol. 35(352), 2010

Tyrrell, John G., 'Site Investigations of Tornado Events', in Robert K. Doe (ed.), *Extreme Weather* (Chichester: John Wiley & Sons Ltd, 2016)

US Census Bureau, '2013–2017 American Community Survey' (Maryland: US Census Bureau, 2018)

Vans, Robert, and St George, 'An Account of an Extraordinary Meteor, or Kind of Dew Resembling Butter, That Fell Last Winter and Spring, in the Provinces of Munster and Leinster, in Ireland; Being Extracts of Two Letters, the One from Mr. Robert Vans to Mr. Henry Million, Dated November 15, 1695. The Other from the Right Reverend St. George, Lord Bishop of Cloyne, to Sir Robert Southwell, U.P.R.S. Dated April 2, 1696. Wherein Mention Is Likewise Made of a Person Having a Regular Epileptick Fit Every Day at a Certain Hour', *Philosophical Transactions of the Royal Society of London*, vol. 19, 1695

Watt, Margo C. and Samantha Di Francescantonio, 'Who's Afraid of the Big Bad Wind? Origins of Severe Weather Phobia', *Journal of Psychopathology and Behavioral Assessment*, vol. 34, 2012

Wakimoto, Roger M. and James W. Wilson, 'Non-supercell Tornadoes', *Monthly Weather Review*, vol. 117, 1989

Wallington, C.E., *Meteorology for Glider Pilots* (London: John Murray, 1961)

Weld, Isaac, *Illustrations of the Scenery of Killarney and the Surrounding Country* (London: Longman, Hurst, Rees, Orme & Carpenter, 1807)

Wilde, Jane Francesca, *Ancient Legends, Mystic Charms and Superstitions of Ireland* (London, Ward & Downey, 1888)

Wilde, Oscar, *The Complete Fairy Tales*, 1st Edition, 1891 (Los Angeles: Norilana Books, 2007)

Xu, You-Lin, *Wind Effects on Cable-Supported Bridges* (Singapore: John Wiley & Sons, 2013)

Yeats, William B., *Fairy and Folk Tales of Ireland* (Oxford: MacMillan Publishing Company, 1983)

Young, Simon and Ceri Houlbrook (eds), *Magical Folk: British and Irish fairies 500 AD to the present* (London: Gibson Square, 2017)

Index

page locators in *italic* refer to illustrations

Abban, St, 83, 199
academic journals, 211
Achill Island, County Mayo, 98
Achill Sound, County Mayo, 106–107
afternoon peak frequency, 130
air pressure, 62, 63, 71. *see also* wind
    speeds
Altan Lough, 109
Anglo-Normans, 36, 37
animals and wildlife, 185–186
Annals
    Book of Leinster, 29
    of Clonmacnoise, 21–23, 24, 37–38,
        204–205
    of the Four Masters, 25, 36, 189, 204
    mirabilia, reports, 198
    tornado events, reports, 34–36
    on turf cutters, 202
    of Ulster, 21–24, 35
    weather and cosmic phenomena,
        reports, 19–20, 204
annual totals of tornadoes and
    waterspouts, 112–114
Antrim, County, 116, 117–118, 153–155
Ardmore/Tolans Point tornado,
    153–155
Ardmurcher House, 143, 145, 146
Armagh, County, 127, 184–185
Armagh Observatory, 85
Athlone, 181
atmospheric pressure, 101

atmospheric systems, and chaotic
    processes, 134–135
atmospheric vortices, 87

Bailieborough, 57
Ballickmoyler tornado, 223
Ballinhassig, County Cork, 195
Ballyclamper tornado, 189
Ballyconnell, County Sligo, 190–191
Ballynahinch, County Down, 192
Ballysadare tornado, County Sligo,
    50–51, 79–81
Banemore, County Kerry, 73–75,
    137–138
Beaufort, Francis, 51–52
Beaufort scale of wind force, 51–52
Belfast Lough, 207
Belfast Newsletter, 206–207
Belfast Telegraph, 208
Belfast tornado, 175, 182
Belgium, 229
Belmullet, meteorological station, 107,
    158
Blacksod Bay, 107–108, 158
Blessington Lake, County Wicklow,
    179
boats, 178–180, 195
boglands, 202–203
Book of Ballymote, 30
Book of Fenagh, 200
Book of Glendalough, 30

Book of Leinster, 29
boundaries, wind direction and speed,
    71–73
Bravo Platform waterspout, 100–102
bridges, 177–178, 215
British Meteorological Office, 212
Broadhaven Bay, County Mayo, 180
Brow Head tornado, 169, 187
building structures, 168–174, 216
    explosive effects, 169–170
    farms, 172–174
building styles, storm-resistant, 46–47
butterfly effect, 135

camera technology, 16. *see also*
    photographs
CAPE values, 72, 75, 76, 98, 100
caravans/caravan sites, 2–3, 176, 189,
    216
Carlingford Lough, County Down, 110
Carraroe, County Galway, 189
Carrickfergus, 207, 208
Carrigallen, County Leitrim, 149–153
Carrigallen tornado, *124*, 125, 130–131
cars, 174–176, 215
Castlederg, County Tyrone, 54
Castlefin, County Donegal, 48
Castleisland, 176, *177*
Cavan, County, 57, 63–64
children
    as eyewitnesses, 221–222
    and PTSD, 9
Christian missionaries, 198
Chronicum Scotorum, 32–33
classic supercells, 78
Clew Bay waterspout, 98–100
climatic chronologies, 131–133
climatic database, 212
climatology, 44, 47–48, 53
Clonbroney, County Longford, 34–36
Clonmacnoise events, 25–29, 30, 200,
    201
Clonmacnoise manuscript, 32–33

cloud forms. *see also* funnel clouds
    cumulonimbus, 6, 59, 68, 93, 149
    cumulus congestus, 69–70, 93, 99
    tornado parent clouds, 68–70
    wall clouds, 69, 78, 101
Cluain-Bronaigh. *see* Clonbroney,
    County Longford
clusters, 126–128
Coalisland, County Tyrone, 12–14
coastal waters and settlements, 86, 91
Colla tornado, 86
computer modelling, 81–82
condensation funnels, 62–63, 64, 76
Connemara, *69*
convective available potential energy
    (CAPE), 72, 75, 76. *see also* CAPE
    values
convent, 34–36
convergence, and vorticity, 91–93, 100
Coolrain tornado, 188
Cork, County, 31, 85–86, 116, 127, 169,
    195
Cork city, 42, 49, 211
Cork Cuverian Society, 42, 211
Cork Examiner, 211
Cork–Wexford coastline, 91
Corraun, 107
cumulonimbus clouds, 6, 59, 68, 93, 149
cumulus congestus, 69–70, 93, 99

damage tracks/patterns, site
    investigations, 150–153, 159–160,
    162–163, 165–166
databases, 228
de Bhaldraithe, Tomas, 39
debris cloud, 61, 64, 67, 155, 186–187
debris fallout, long-distance, 189–193
debris field, 162–163, 187–189
demographic change, 219–220
Department of Transport, 45
Derry Water, 192
Derrynacross, County Longford,
    54–55, 176, 183–184

dictionaries, 38–39
directional wind shear. *see* wind shear, directional
Doagh Isle, County Donegal, 109
Domnall mac Murchada, King, 29
Donaldson, Joseph, 55
Donegal, County, 108–109, 117, 201
Down, County, 110, 127, 192
downbursts, 165
downdraft, 77, 101, 190
dragons, 21–24, 199, 200
Dublin Bay tornado, 147–149
Dublin Evening Mail, 43
Dublin tornado, 42–43, 182, 210
Dunfanaghy, County Donegal, 201
Dungarvan tornadoes, 133–134
Dungarvan's Ballyclamper Holiday Park, 189
Dunlavin, County Wicklow, 191
Dunlewey Lough, 108–109
dust devils, 196

economic impact, 47
eddy waterspouts, 106–108, 110
Ehrenberg, C.G., 204
Einstein, Albert, 16
emergency workers, and PTSD, 9
emotional responses, 8, 9–10, 137, 196, 212, 213
environments, Irish, 138, 202–203
ESSL (European Severe Storms Laboratory), 228
ESWD (European Severe Weather Database), 228
Europe, 228–231
eyewitnesses, 55, 73, 98, 144–145, 211. *see also* metaphors; *specific names of individuals*
    children as, 221–222
    emotional responses, 8, 9–10
    interviews/questionnaires, 163–164
    major source of information, 167

and social/cultural attitudes, 15–16, 51
technologies for confirming phenomenon, 58

facial deformity, 197
faeries, 196
faery wind/blast, 39–40, 195, 197
fair weather waterspouts, 90, 93, 97
fake images, 11
farm buildings, 172–174
Farm Buildings Scheme, 173
fatalities, 86, 97, 196, 200
Fennor (Finnabhair), 193
field observations, 137
Finley, John, 224
Finn Valley Voice, 48
fish, 189, 190–192
fishing boats, 105, 195
Florida coast, 97
folklore tales, 197–198
food production, and health, 43
footprint, tornado, 137–149
forecast experiment, Bravo Platform waterspout, 100–102
forecasting tornadoes, 45, 54–57. *see also* TORRO
France, 46
frictional drag, coastal, 91
Fujita, Theodore, 69
Fujita Scale (F-Scale), 52
funnel clouds, 50, 56, 59–60, 64, 94–95, 147
    seasonal variation, 128–129
    winter and summer, 122–123

Galty Mountains, County Tipperary, 64–66
Galway, County, 66, 96–97, 116, 189
    and Roscommon, tornadoes, 138–141
Germany, 229
Gilbert, Tony (forecaster), 141

Glenflesk, County Kerry, 84–85
Glenveagh National Park, 108
Gosford Forest Park, 184–185
grass and peat, economic crops, 47
Great Famine, 1846–51, 43
Great Lakes, USA, 97
Great Plains, USA, 226
Griffin, Dr Daniel (of Limerick), 42,
   170, 210

hail, 44, 78, 79
Hanrahan, Dick, 133
hazardous material, debris fallout,
   189–190
high precipitation (HP) supercells, 78
history of Irish tornadoes
   earliest records, 17–20
   metaphors, 20–34
Hogan, Dick, 143
homes, damage to, 171–172
hook echo, 77, 78
horizontal spin (vorticity), 90. *see also*
   convergence, and vorticity
Horseleap tornado, 146
Horseleap/Streamstown area, 143, 145,
   146
Hynes, Linda, 67–68

impacts of tornado, study of, 137. *see*
   also damage tracks
Inishowen, County Donegal, 109
intensity of tornado, 115, 125–126,
   158–159
International Meteorological Society,
   44
interviews/questionnaires, 163–164
investigation. *see* site investigations
Irish Air Corps, Maritime Squadron,
   98
Irish Annals. see Annals
Irish Coast Guard (IRCG) helicopter
   rescue unit, 59–60
Irish Examiner, 1

Irish Language
   for waterspouts, 84
   for whirlwinds and tornadoes,
      38–40, 84
Irish Meteorological Service, 45. *see*
   also Met Éireann
Irish Naturalists' Journal, 192
Irish Nennius, 30
Irish Times, 1
Island Magee, 207, 208
islands, offshore, 86

jet stream, 92, 226

Kerry, County, 47, 59–61, 73–75,
   84–85, 137–138, 180
Kerry Airport, 176
'kettles,' 110
Kilbeggan, County Westmeath, 32,
   33–34, 145, 201–202
Kilbeggan Secondary School, 145
Kildare, County, 174
Kilkenny, faery blast, 195
Killeshandra, County Cavan, 63–64
Kilmallock, 181
Kilroosky, County Roscommon, 138,
   140
Kinrush funnel cloud, 13–14
Kippure Mountain, County Wicklow,
   175
Kongs Skuggsjo, 20, 25–26

land breeze, 92
land surface temperatures, 88–89
Lane, Timothy O'Neill, 38–39
language. *see also* Irish Language
   vocabulary for whirlwinds/
      tornadoes, 36–40
   vocabulary of denial, 48–49, 208
Latin, 36
Lecarrow, County Roscommon, 138,
   140
Leitrim, County, 125, 149–153, 200

Limerick Chronicle, 42
Limerick tornado, 42, 43, 170, 181
Lloyd, Rev. Humphrey, 210
   1851 survey, 41–44
Logh Kynn, County Galway, 37
long-distance debris fallout, 189–193
Longford, County, 34–36, 183–184
Lorenz, Edward, 135
Lough Bane, 87
Lough Beagh, 108
Lough Fenagh, *199*, 200
Lough Hacket, 37
Lough Neagh, 84, 87, 118
   waterspouts, 102–103
loughs, and weather hazards, 87
Louth, County, 116
low precipitation (LP) supercells, 78, 79
Lucretius, *De Rerum Natura*, 83
Lynch, Dan, 65

MacDonald, Daniel, *Sidhe Gaoithe/The Fairy Blast*, 40
Markethill tornado, County Armagh, 184–185
Massachusetts, USA, 224
Mayo, County, 106–108, 116, 158, 180
McDonald, Conor, 12
McKenna, Lambert, 39
McKenna, Martin, 12
McSherry, Brendan, 65
Meath, County, 67–68, 127, 193
media
   emphasis on exceptionalism of tornadoes, 206, 208
   landmark footage, 50
Meelick, County Galway, 138–139
mesocyclone, 76, 77, 78, 79–80
mesoscale weather events, 44
Met Éireann, 16, 48, 50, 53–54, 212
metaphors, 20–34
meteorological environment, data, 165
meteorological instruments, 41, 101, 224

meteorological research, 41–44
meteorological stations, 158, 209
meteorology, 44, 53
Meyer, translation of *Kongs Skuggsjo*, 25–26
Michigan Argus, 205
mirabilia, 198, 200, 203–205
mobile homes, 188–189, 216. *see also* caravans/caravan sites
moist convection, 72, 76
monastic chronicles/culture, 19, 36. *see also Annals*
monthly patterns, 121–131
Mount Errigal, 108–109
mountain areas, 119–121
Mullingar tornado, 143, 145, 181
multiple cell events, 67
multiple vortices, 146

National Oceanic and Atmospheric Administration (NOAA), 225, 227
near misses, 176–177, 199
Nenagh, County Tipperary, 42, 211
newspapers. *see* press
night tornadoes, 130–131
Nolan, Mr and Mrs, 148
non-supercell tornadoes, 70, 166
Norse, 20
North America. *see* USA
nozzle effect, 66
nuns, 34–36

Ó Síocháin, Conchúr, 40
O'Connor, Hugh, 224
O'Donovan, John, 85
O'Driscoll, John and Jack, 85–86
O'Dwyer family, 189
Offaly, County, 127
offshore islands, 86
O'Hegan family, 14
Oklahoma, tornado forecast, 54
Old Norse, 36, 37–38, 110
oral sources, 20

O'Shea, Brian, 158
O'Sullivan, Humphrey, *Diary of an
    Irish Countryman, 1827–1835*, 38,
    84–85

Patrick, Bishop of Dublin, *Bishop
    Patrick's Poem*, 31
peat extraction, 215
Peters, Michael, 68
*Philosophical Transactions of the Royal
    Society*, 205
phobias, 9
photographs, 11, 16, 166–167
popular story telling, 196
population density, mountain areas, 121
posttraumatic stress disorder (PTSD),
    9, 14, 220
poverty, 219
Preban, County Wicklow, 54
press. *see also names of specific
        newspapers*
    local, tornado reports, 49
    urban, 206–208
    vocabulary of denial, 48, 49, 208
psychological issues. *see emotional
        responses*
public, vulnerability of, 211, 214–216
public survey, 1999, 51

questionnaires, 164
Quirke family, 189

radar technology, 78, 81, 137
rail network, 176–177
rainfall observations, published
        reports, 209
Rathangan, County Kildare, 174
rear flank downdraft, 77
Reeves, Bishop William, 83–84
refuges, 214–216
regional climatologies, 44
regional variation, 116–119
remote sensing images, 165

road networks, 177
Roaringwater Bay waterspout,
        85–86
Robertstown, County Meath, 67–68
roof structures, large span, 172–174
Roscommon, County, 138–141
Rosemount tornado, 146
Ross-deala, 32–34
Roundstone, County Galway, 66
Roundstone waterspout, 96–97
rowdows, 105
Royal Irish Academy (RIA)
    1851 survey, 41–44
    established, 41
    role of, 209–210
Royal Society of London, 209
rural areas
    and damage, 161, 172–173
    population, 46
Ryanair, 176, *177*

scientific observation, 18
sea surface temperatures (SSTs),
        88–90
seasonal and monthly patterns,
        121–131
Severe Thunderstorm Watch, 54
severe weather phobia, 9–10
Shakespeare, William, *Antony and
        Cleopatra*, 18
Shannon, 59–60
Shields, Lucie, 12
ships in the sky (metaphor), 24–32,
        200–201
shower events, 203–205
sidhe gaoithe, 39–40, 45, 194–198, 221
simulations, laboratory, 159, 175
site investigations, 136–167, 143–145
    data gathering, 159–167
Slane, County Meath, 36, 193
Sleater, Keith, 50
Sligo, County, 50–51, 79–81, 190–191
social and cultural attitudes, 15–16, 57

sounds of tornadoes, 10–11, 197, 216–218
specialist magazines, 45
St Patrick's Day tornado, 141
Stack Hills tornado, 59–61
statistical probabilities, 132–133, 134
sticklebacks, 192
stone crosses, 36
storm cell waterspouts, 93
storm chase method, 55–56, 79, 225
storm narratives, 221
Streamstown, 143, 145
stretching mechanism, 71–72, 93–94, 191
suction vortices, 67
supercell tornadoes, 70, 75–81
surface convergence, airflow, 100
surface friction, 87–88
surface vortices, 191
surface water temperature, 88–90
surface-based triggers, and waterspouts, 98
surface-level water vapour, 90
Swift, Jonathan, 18
Symons's Monthly Weather Magazine, 209
synoptic charts, 6, 147–148
synoptic stations, 53–54

Taillten (*now* Teltown), County Meath, 29–31, 193
Tatlock, John S.P., 204
Temperley, Hans, 143
Termonamongan, 208
Texas, Oklahoma, 224
Thetis, HM Packet, 180
thunderstorms, 44, 65, 70, 101, 166
timber frame housing, 172
Tipperary, County, 56, 64–66, 211
Tirconnell, fish fall, 189
Tolans Point, 153–155
topographic associations, 119–121
tornadic waterspouts, 97

tornado, elements of a, 61–70, 81–82
appearance, 61–63
inflows, 67–68
parent clouds, 68–70
single and multiple cells at surface, 66–67
surface processes, 63–66
Tornado Alley (2011), 111
tornado alley, Ireland, 116–117
Tornado Alley, USA, 111, 224–226, 232
tornado clusters, 126–128, 232
tornado days, 114–115, 128
Tornado Research Organisation (TORRO). *see* TORRO
tornado totals, Europe, 228–229
tornado tracks, 33. *see also* damage tracks/patterns
and intensity, 158–159
length of, 153–156, 159, 193
life cycle, 149–153
and site investigations, 136–137, 143–146, 153–159
width of, 156–158, 159
torrential rain, 67, 68, 85, 155
TORRO, 52, 54, 56, 100, 141–142, 228, 230
tower of fire (metaphor), 32–34, 201–202
transport systems, 176–178
trauma, 137, 220
tree damage, 183–185
troposphere, 98
T-Scale, 52
Tumona event, 202–203
turf cutters, 202–203
Twister (1996), 111
Tyrone, County, 12, 54, *188*, 208

UK, 46, 228, 230
University College Cork, 16, 52, 112
urban areas and structures, 181–183
Urney, County Tyrone, 208

USA, 223–227
    CAPE values through troposphere,
        98
    estimates on supercell/non-supercell
        events, 70
    forecasting tornadoes, 54
    and Irish ancestry, 223–224
    Midwest, 12, 97, 111
    research on CAPE values, 72
    tornado database, 226
    tornado events, reports, 46
    waterspout research, 97

Valentia meteorological station, 101,
        120, 142
vehicles, 174–176, 177, 215
verification, 136–137
vernacular, annalists use of, 36
vertical stretching, 93–94, 101
Vikings, 20, 36, 37, 202
voluntary research organisations, 160.
    see also TORRO
vorticity, 90. *see also* convergence, and
        vorticity
vulnerable people, 218–222

wake vortices, 92, 120
wall clouds, 69, 78, 101
water bodies, 178–180
water monsters, 83–84
water surfaces, and friction,
        87–88
water vapour, surface-level, 90
Waterford, County, *33*,
        133–134

waterspouts, 37, 83–110
    environments, 90–97, 178–180
    and long-distance debris fallout,
        190–192
    vocabulary, 39
weather enthusiasts, 53, 160
Western People, 192
Westmeath, County, 67, 68, 116, 127
Westmeath tornadoes, 141–146
Wexford, County
    South East Vegetables, 188
    waterspout event, 83, 199
Wexford tornado, 49, 181
whirlwind events, 196–198
whirlwinds and tornadoes, vocabulary,
        36–40
Wicklow, County, 116, 175, 179,
        191–192
Wilde, Lady, 195
Wilde, Oscar, 195
wind shear, 76, 80, 92, 98, 100, 108, 142,
        183, 191
wind speeds, 67, 171–172, 173. *see also*
        convergence, and vorticity
        boundaries, 71–73
        and directional change, 80–81
Wokingham tornado (UK), 228
wooden crosses, 36, 192–193
woodlands, 183–185
Woodworth, Simon, 65
World Meteorological Organization
    (WMO), 44

Yeats, William Butler, 197
Youghal, County Cork, 1–7, 48